Vacant Possession: Law and Practice

T0340322

For the Leeds PLT

True Property Warriors!

Vacant Possession: Law and Practice

Keith Shaw

Routledge
Taylor & Francis Group

LONDON AND NEW YORK

First published 2010 by Estates Gazette
First edition 2010

Published 2014 by Routledge
2 Park Square, Milton Park, Abingdon, Oxon OX14 4RN
711 Third Avenue, New York, NY 10017, USA

Routledge is an imprint of Taylor & Francis Group, an informa business

British Library Cataloguing-in-Publication Data
A catalogue record for this book is available from the British Library

Library of Congress Cataloging-in-Publication Data
A catalog record for this book is available from the Library of Congress

ISBN: 978-0-08-096680-9 (pbk)

Contents

Foreword

Vacant possession is a concept which lawyers, surveyors and others concerned with land and buildings frequently come across in practice, in both freehold and leasehold contexts. However, despite its importance, the nature and meaning of vacant possession has, up to now, been dealt with in various textbooks and handbooks in a cursory and limited way. This is unfortunate and surprising because vacant possession is, as Keith Shaw says in this book, a pervasive concept, and it raises a number of interesting and difficult issues, which are often overlooked or confused, even by those with experience and expertise in the property world.

Vacant possession is also a rather imprecise expression, and, as this book demonstrates, the judiciary has not always helped, in that the case law does not display a clear, consistent or principled approach to the meaning or effect of the expression. However, to be fair to the judiciary, as with virtually every word, sentence or expression, the precise meaning is very much governed by its particular context. Further, there has been a surprising scarcity of cases (or at least of reported cases) on what vacant possession actually entails, so the opportunity to develop a satisfactory jurisprudence on the topic has been rather lacking. This seems to me to render it all the more surprising that it is rarely, if ever, that a contract, whether standard or bespoke, actually contains a definition of vacant possession, even now, when almost every contract has many pages of definitions.

Thanks to this book, there is no longer any excuse for such ignorance and confusion. For the reasons discussed above, it fills a little noticed, but very significant, gap in the market. The book should be of great value in that it will, I hope, ensure that conveyancers, litigators, surveyors, estate agents and others, including property owners, landlords and tenants, concerned with property are fully and properly informed about the law and practice relating to vacant possession. Over and above that, this book draws attention to the importance of parties to a contract directing their minds to what they mean by the expression when they use it in a contract.

More specifically, like any good book on an apparently narrow subject, this casts a much wider net than one might expect from the first impression made by the title. Thus, quite apart from difficult questions such as the meaning of possession (a slippery concept), Keith Shaw considers the law relating to implied terms, the history of the standard conditions of sale, the effect of compulsory purchase orders on sale contracts, the law relating to fixtures, actual state and condition clauses, and the law relating to topics such as completion, break clauses and rent review clauses. I have always found it useful and informative to read about legal concepts and principles from a specific perspective, as opposed to reading about them in a more generalised context. Thus, quite apart from its inherent value in examining and explaining the concept of vacant possession, this book represents a valuable contribution to a number of other important aspects of property law.

Any book about law, especially one concerned with a relatively ancient and technical concept which is of very general application, and which often has important and sometimes wide-ranging practical consequences, should be easy to use and accessibly written. I am glad to be able to say that Keith Shaw has written a book which is sensibly and practically structured, and well and clearly written. All those who work in the property world should be grateful to him.

The Rt Hon Lord Neuberger of Abbotsbury
Master of the Rolls
President of the Property Litigation Association
April 2010

Preface

This publication arose out of an article entitled 'How Vacant is Vacant Possession?', which was originally submitted to the Property Litigation Association for its 'essay writing' competition in November 2006 and subsequently published by the *Estates Gazette* as 'Fit to be Occupied' in January 2007.[1] This article questioned the nature, scope and extent of the obligation to give vacant possession, something generally understood as a commonly used and rather simplistic property law concept.

It is correct to say that a wide range of people use the expression 'vacant possession'. Indeed, estate agents even seem to be able to distinguish, in their advertising particulars, between 'full vacant possession', 'immediate vacant possession' or 'complete vacant possession'; there is ostensibly no real justification as to how the preceding adjective in each case adds anything to the message that they are seeking to convey to prospective purchasers, as to what they can expect to obtain on completion. Lawyers talk about 'giving VP on completion', but few documents ever actually define what vacant possession means with a capitalised 'V' and 'P'.[2]

Whilst an everyday term that is used by many, behind the familiarity of this common expression one can uncover hundreds of years of uncertainty, misunderstanding and general neglect of the development of a sound and coherent theoretical model of vacant possession. There is very little case law and even less judicial guidance available. In 1988, and in two editions of the *Conveyancer and Property Lawyer*, the barrister Charles Harpum wrote what probably remains the most insightful learned article on the subject,[3] but since then the concept appears to have warranted very little scholarly or practitioner attention.

This book follows three years of doctoral research undertaken through Durham University into the concept of vacant possession and its meaning. Based on the thesis, the book is specially aimed at legal practitioners (of which the author is one) and other professionals and individuals who encounter the term 'vacant possession' on a regular basis. The book outlines the current law surrounding the obligation to give vacant possession, along with the practical application of the term in real life scenarios. In doing so, it highlights the difficulties that lawyers, surveyors and third parties face on a day-to-day basis when seeking to interpret the nature, scope and extent of the obligation.

---

[1] See Shaw, K. 'Fit to be Occupied' (2007) *Estates Gazette* 182–183.

[2] See Shaw, K. 'Bone of Contention' (2009) *Estates Gazette* 58–59.

[3] Harpum, C 'Vacant possession — chamaeleon or chimaera?' (1988) *Conveyancer and Property Lawyer* 324, 400 (CH).

In the recent economic decline, parties have more than ever before sought to find novel (or convenient) ways to seek to avoid payments and liabilities under contractual documents and other agreements. An exploration of this obscure concept, as a means of seeking to frustrate legal obligations at a given time, is therefore both topical and timely in the current economic climate. In particular, the use of the obligation to give vacant possession in the leasehold context (and with reference to a tenant's compliance with conditional break options) is shown to be an area where the obligation's opaque nature can be used to gain tactical advantages in litigation.

To the everyday man or woman on the street, the term 'vacant possession' raises its head most noticeably in the residential sphere, with many everyday people buying and selling property only to find that their obligation to give, or right to receive, vacant possession is far from straightforward. Still to this day, the vast majority of residential sale and purchase contracts are completed *before* an inspection of the property to which the contract relates, with the completion monies thus having been sent over to the seller in advance of the purchaser being able to confirm that it has received vacant possession.

This book explains why a working understanding of the term is essential to practitioners and others who encounter issues of vacant possession, and how a lack of proper consideration of the meaning and implications of the term can lead to significant legal problems down the line; something many have already experienced to their cost and inconvenience!

A great number of people have assisted me in the preparation of this book, including many colleagues, family and friends. It would be rather long winded to recite them all but it is appropriate, in particular, to thank for their assistance, colleagues at Pinsent Masons LLP (including, in particular, Stuart Wortley, who originally suggested that the subject of vacant possession 'would be worth looking at for the PLA essay competition, as not much has been written about it'), Durham University (my PhD supervisors, Professor Lorna Fox-O'Mahony and Neil Cobb, who have enabled my understanding of the subject to become more scholarly and informed) and Katherine Holland QC of Landmark Chambers (who read some of my early work on the subject and also draft chapters of the manuscript). I am also indebted to those who proofread the first draft of this publication, and thank Alicia Foo and Alister Bould. To single one person out: a great debt is owed to John Martin (formerly of Pinsent Masons LLP), who enabled this undertaking to get off the ground from the beginning; his support and assistance were most valuable to me, and I wish him well in the future.

Keith Shaw
Solicitor, Property Litigation Department, Pinsent Masons LLP
Researcher, Durham University Law School
Alexander Maxwell Law Scholarship Trust Award Winner – 2009

Alexander Maxwell Law Scholarship Trust

LAW SCHOLARSHIP TRUST

Maurice W. Maxwell, whose family founded Sweet & Maxwell the law publishers, by his Will established a charitable trust to be known as the Alexander Maxwell Law Scholarship Trust in memory of both his great-great-grandfather and great-grandfather.

The Trust is committed to promoting legal research and writing at various levels via the provision of financial assistance to authors involved in legal practice, whether they be experienced legal practitioners, or those in the early years of practice. (Please note that the Trust does not make grants towards the cost of any educational or vocational courses or post-graduate work.)

The author of this work received financial assistance from the Trust to enable him to complete the work.

The Trust calls for applications for awards each year, with a closing date in the preceding Autumn (see website for details and application forms and other guidance on submission). Applications are reviewed by the Trustees, with the assistance of Sweet & Maxwell; shortlisted candidates are interviewed by the Trustees and awards are made shortly afterwards.

Anyone interested in its work should write to:

The Clerk to the Trustees
Alexander Maxwell Law Scholarship Trust
c/o Sweet & Maxwell
100 Avenue Road
London NW3 3PF
e-mail: amlst@thomson.com

The Trust's website is at: http://www.amlst.org.uk

Chapter 1

What is Vacant Possession?

Chapter Outline

The obligation to procure vacant possession is generally understood to refer to the legal commitment to ensure that at the relevant date (for example, on completion of contracts or termination of a lease) a given property is in a state fit to be occupied **1**

Vacant Possession: Law and Practice. ISBN: 978-0-08-096680-9

(both physically and legally) and enjoyed.[1] Vacant possession is known to be relevant to the sale of freehold land and property (e.g. the transfer of estates in fee simple) and upon the grant, transfer and termination of leases and other tenancies. Vacant possession is an essential element of any land transaction where the right to occupy a property is being vested in, or passed to, a third party. The only category of transaction for which vacant possession will not be relevant concerns the purchase of reversionary interests (for example, a freehold subject to a long lease) and other estates that are not 'vested in possession'.

The term vacant possession is often, quite curiously and perhaps inappropriately, used with respect to informal agreements to occupy by consent. Upon termination of a licence a tenant will normally be obliged to vacate the premises, but it is arguable as to whether this is giving vacant possession as the tenant is not returning possession (unlike, for example, upon the termination of a lease). The essence of a licence, as opposed to a lease, is that it does not amount to a legal estate or interest in land and does not bind third parties, but it is nevertheless common for licences to (wrongly) make use of the term 'vacant possession' (rather than 'vacating') in this context.

Figures obtained from the Land Registry Annual Reports and Accounts 2008/09 confirm that in the 255 working days of the year to which these accounts referred, there were over 319,000 first registrations (comprising land conveyed or leases granted out of unregistered titles). There were over 198,000 leases granted out of existing registered titles during this period. There were around four million dealings with registered titles (the transfer of whole or part of an existing registered title). These types of transactions will have involved the issue of vacant possession. That is nearly five million transactions each year, some 20,000 transactions per working day, 3,000 transactions an hour and therefore approximately 60 transactions every minute. In other words, a transaction to which vacant possession will be relevant takes place the equivalent of every second of each working day in England and Wales.[2]

It is therefore unsurprising that vacant possession is one of the most commonly used terms in the property professional's dictionary. Given that vacant possession arises on the majority of land transactions in England and Wales, the sheer scale on which the obligation is engaged causes the term to be relevant to a wide range of people. These include sellers and purchasers, landlord and tenants and third party occupiers both on residential and commercial transactions. From the outset, it is important to appreciate why vacant possession is something of interest, and concern, to such people.

---

[1] See Shaw, K 'Fit to be occupied' (2007) *Estates Gazette* 182–183.

[2] See the Land Registry Annual Report 2008/09. Calculation based on a working day comprising 6.5 working hours.

Vacant possession and people

Vacant possession is key to the majority of land transactions in a residential and commercial context and presents a significant area of risk for sellers and buyers, and landlords and tenants.

RESIDENTIAL AND COMMERCIAL TRANSFERS

On the sale of a residential property, a seller will be keen to ensure that vacant possession is provided; otherwise the buyer may refuse to complete the purchase. If the buyer refused to complete because the premises could not be occupied without difficulty or objection on completion, the buyer could serve a notice to complete, and ultimately rescind the contract thereafter if the seller does not, or cannot, give vacant possession. All this will concern a seller who may have exchanged and be seeking completion themselves on the related purchase of a new property which he or she will not be able to finalise until the sale of their current property has completed. On the purchase of the new property, the seller (as buyer in respect of that purchase) could lose his or her deposit if unable to complete on time because the sale of an existing property has not taken place, causing the seller not to have the funds required to complete his or her proposed purchase. It is known for people in such circumstances to ultimately end up homeless in the interim when difficulties in completing the sale of their current property cause them to lose the right to purchase the property for which they have contracted to buy.

If the buyer does complete, only to then find that vacant possession has not been given, this would leave the buyer entitled to damages as a consequence of the seller's default. Such damages could be quite significant and impose a financial burden to the seller. A seller is therefore well advised to ensure that the premises are fit for occupation by the buyer on completion, irrespective of whether the seller has a related purchase to complete.

Normally on the sale of a residential property the parties will complete the transaction *before* an inspection of the property has taken place. It is suggested that ideally 'the buyer's conveyancers should check for any evidence as to rights of occupiers by either personally inspecting the property or advising the buyer client to do so…the buyer's conveyancers should raise a requisition of the seller's conveyancers requesting confirmation that vacant possession of the whole of the premises will be given on completion and that all occupiers have agreed to vacate'.[3] In practice this does not normally occur. The fact that the majority of residential transactions are

---

[3] See Lexis Nexis Butterworths Document [547] 10 Occupiers. Accessible by subscriber service.

completed *without* an inspection immediately prior to completion, poses risks for all parties involved.[4]

The difficulties caused by not giving vacant possession can be even more onerous in the commercial context. A buyer may be a commercial organisation requiring immediate unencumbered possession on completion in order to ensure business continuity from their previous premises. If vacant possession is not given on completion, then putting aside the legal remedies available, the buyer could lose important contracts or strain business relations and goodwill with existing suppliers and trades people because of its inability to continue to trade normally. Commercial organisations are often tied to service and other contracts which will include fixed penalties for non-compliance in accordance with contractual agreements. The consequences of not being able to immediately occupy a given property (or set of premises) can therefore be very serious.

Leases

The most common situations in which vacant possession is relevant in the commercial context are with respect to the grant, transfer and termination of business leases over properties.

Grant and transfer

If the lease is to be 'vested in possession' it will be implied (as part of the grant) that vacant possession is being given. If the lease was to be vested only 'in interest' (i.e. a reversionary lease that is granted to begin at some time in the future, usually after the prior existing lease has expired), then vacant possession would not be implied as it would not be intended to give an immediate right to enjoy the estate in land.[5] The right to enjoy the estate, in such case, would be postponed to some future date, when its term would start.

If a tenant is granted a lease only to find that it cannot take up occupation, then that tenant would likely advance a claim against the landlord for derogation from grant and/or a breach of the landlord's express or implied covenant for quiet enjoyment.

If the lease is being granted pursuant to an agreement for lease (perhaps conditional on such matters as planning permission being granted or landlord's works), then the agreement for lease is likely to incorporate (by reference) the *Standard*

---

[4] See chapter 8 for a discussion of the need to time the moment of completion with the procurement of vacant possession.

[5] See *Long* v *Tower Hamlets LBC* [1996] 2 All ER 683.

Commercial Property Conditions (second edition); unless amended, these will include the giving of vacant possession upon the grant of the leasehold interest as a special condition.[6]

In similar terms, on the assignment of a leasehold interest it will be implied that the assignee will be able to take possession of the property pursuant to the assignment. If the assignment is pursuant to an agreement to assign, this may also incorporate (by reference) the *Standard Commercial Property Conditions* (second edition) which will similarly include vacant possession as a special condition, unless an amendment to the conditions is made. The implication arises because there is not usually a contract on the assignment of a lease, and vacant possession will often not be dealt with expressly, unlike in the context of the sale and purchase of freehold land.

Termination by break option

It is common for a tenant to wish to bring their lease to an end before its contractual expiry by exercise of a 'break option'. Commonly, exercise of the break is conditional on the procurement of vacant possession, or compliance with all covenants in the lease up to and including the break date. If the lease includes a covenant to yield-up the premises on termination, this will include an obligation to give some form of vacant possession in any event, even if the procurement of vacant possession is not expressed to be a pre-condition for operation of the break in its own right.[7] This is particularly a problem given that:

> there has been very little guidance from case law as to what constitutes yielding-up...[n]either has Parliament ever prescribed the meaning of the expression.[8]

Therefore, this creates uncertainty for all parties involved.

In a changing, and recently downward, property market, it is common for tenants to seek to reduce operating overheads and source alternative (and normally smaller) premises. This can only be achieved if their current lease is brought to an end. Giving vacant possession is therefore essential for the tenant in order to ensure that its current lease does not continue past the break date. If the tenant fails to operate the break option successfully and remains tied to its current lease along with a new tenancy that may already have commenced, or the tenant becomes contractually bound to enter into, this could lead to dire financial circumstances. It is common for tenants to misunderstand the requirements for a successful operation of a break

---

[6] See chapter 4 for a discussion of the *Standard Conditions of Sale* (London, The Law Society 4th ed, 2003) and the *Standard Commercial Property Conditions* (London, The Law Society, 2nd ed, 2003).

[7] See chapter 9 for a discussion of how yielding-up will encompass a form of vacant possession.

[8] See Higgs, R 'Leave Your Keys on Your Way Out' (2005) 155 *New Law Journal* 149.

option and unwittingly find that they remain liable and tied to the covenants contained in their current lease because, amongst other conditions, vacant possession was not given at the material time.

If a tenant wishes to operate a contractual break option in a lease, the financial or covenant strength of the tenant are relevant considerations for the landlord in such a situation. If the landlord has a tenant in occupation of its premises under the terms of a lease with 'good covenant strength', or the tenant is paying rent higher than current market value, then there can be little incentive in allowing the tenant to move on. This may leave the landlord having to find a new tenant or face paying empty rates on a property earning no revenue in the short term.[9] The bespoke nature of some premises can cause the potential for an open market letting to be very restrictive in respect of appealing to only a limited range of potential new tenants (for example, an industrial premises for the production of certain toxic chemicals). Further, commonly to secure a new letting, a landlord may have to offer other financial incentives (such as rent free periods) to potential new tenants, which can be a financial strain to the landlord. All this work would cause the landlord to incur time and cost in any event, and is therefore neither appealing nor attractive.

As a consequence, a landlord may wish to prevent a tenant successfully exercising a break option in such circumstances, and use whatever arguments are available to it to claim that the lease has not come to an end. Martin[10] explains that common pre-conditions to exercise a tenant's break option include payment of all rent(s) due up to and including the break date (which itself can be an issue when the break date falls on a quarter day, upon which the whole quarter's rent falls due in advance) and vacant possession. Perhaps the most effective and decisive means of preventing a lease coming to an end, as per the tenant's intention, is if the landlord can construct an argument to the effect that vacant possession was not given at the material time. This can be because of a physical or legal impediment preventing enjoyment of the property at the relevant date. If this proves to be the case, a landlord may thereby force a tenant to remain principally liable under the terms of the lease with the corresponding ongoing liabilities (including, most importantly, the payment of rent).

If a tenant fails to operate a break option successfully then, in the absence of agreeing a surrender of the lease (for which the landlord is likely to charge a premium), it may only be able to divest itself of its ongoing liabilities by assigning or subletting the lease to a third party. Whether and how the tenant can do this will be determined by the terms of the lease and relevant statute.[11] As such, the

---

[9] See Todd, L 'Empty Property Rate Relief – Have the fears become a reality?' (2008) 217 *Property Law Journal* 6–7.

[10] Martin, J 'Tenant's Break Options' (2003) 153 *New Law Journal* 759.

[11] For example, the statutory provisions of the Landlord and Tenant Act 1927.

consequences of not successfully operating a break right, because of failing to give vacant possession, can be potentially very significant, as explained in more detail in chapter 9.

Expiry by effluxion of time

Vacant possession is also relevant to the return of possession when a lease comes to an end by effluxion of time (and the tenant has no statutory or common law rights to remain in occupation of the property).[12] It is common for landlords to advance damages claims when a tenant does not give vacant possession on the contractual or statutory termination date which causes loss or inconvenience as a consequence. Normally, such a claim will be part of other so-called 'terminal' claims including claims for dilapidations (as a consequence of the tenant's failure to keep the premises in the state of repair required by the covenants contained in the lease).

A PERVASIVE ISSUE

Vacant possession is therefore key to the majority of land transactions in a residential and commercial context and presents a significant area of risk for sellers and buyers, and landlords and tenants. It can be an issue when, innocently, two people agree to the sale and purchase of a property and something prevents the premises being fit for occupation at the relevant time. In contrast, it can be invoked as a 'commercial sword' when landlords seek to use the vacant possession obligation to protect their financial position by preventing a tenant bringing a lease to an end prematurely. As the issue of vacant possession has a real impact on people and their lives, a working understanding of the obligation is essential in order to appreciate the responsibilities and liabilities of the parties involved, and the issues of risk which arise in transactions involving vacant possession.

What is the obligation?

In seeking to understand the obligation to give vacant possession, a starting point is to understand what form the obligation will take. The obligation to give vacant possession will normally appear as a 'term' in a legal agreement, conveyance, contract or transfer. This can either be express or implied.

---

[12] See Bowes, C and Shaw, K 'Time's up...but I'm staying!' (2008) 218 *Property Law Journal* 9–11 and Bowes, C and Shaw, K 'Term of years...uncertain' (2009) 225 *Property Law Journal* 7–8 in respect of possession on the termination of leases.

Express Terms

Express terms are terms that have been specifically mentioned and agreed by both parties at the time the contract is made.[13] They can be oral but will usually be in writing as part of the legal contract or agreement. It is essential for express terms to appear in writing in the contract or legal agreement for the disposition of an interest in land in order for there to be a valid contract under section 2 of the Law of Property (Miscellaneous Provisions) Act 1989.

Implied Terms

It is common, however, for a term which has not been mentioned by either party to nevertheless be incorporated as part of a contract, because the term is necessary for the contract to have commercial sense. These terms are known as implied terms and can be implied by either statute (for example, various implied terms in the Sales of Goods Act 1979) or the courts. When implied by the courts this can be as a matter of fact or as a matter of law.

Implied terms of fact

When implying terms of fact, the exercise involved is that of ascertaining the presumed intention of both parties to the contract collected from the words of the agreement and the surrounding circumstances.[14] Something that is so obviously included that it did not need to be mentioned in the contract will normally be implied as a matter of fact. For example, if a purchaser agrees to buy a car from a seller for 1,000 in Whitby, North Yorkshire, it will be implied as a matter of fact that the parties will be referring to pounds Sterling and not US Dollars. Indeed, it has been said that:

> *Prima facie that which in any contract is left to be implied and need not be expressed is something so obvious that it goes without saying; so that, if while the parties were making their bargain an officious bystander were to suggest some express provision for it in the agreement, they would testily suppress him with a common 'Oh, of course'.*[15]

Implied terms of law

Terms can also be implied by the courts as a matter of law, and this involves more general considerations of public policy where the courts are prescribing how the

---

[13] Furmston, MP *The Law of Contract* (London, Butterworths Tolley, 3rd ed, 2007) section 3.1.

[14] See *Ali* v *Christian Salvesen Food Services Ltd* [1997] 1 All ER 721, per Waite LJ at 726.

[15] *Shirlaw* v *Southern Foundaries Ltd* [1939] 2 KB 206, per Mackinnon LJ at 227.

parties to certain types of contract 'ought' to behave. It is said that 'terms implied in law are in reality incidents attached to standardised contractual relationships'.[16] Terms are implied in law where the contract is of a defined type, encompassing:

> *those relationships which are of common occurrence, such as...seller and buyer, owner and hirer, master and servant, landlord and tenant, carrier by land or by sea, contracts for building work and so forth.*[17]

The implication is not based on the parties' intention 'but on more general considerations'.[18] There are two basic requirements for the implication of a term in law:

> *the first requirement is that the contract in question should be of a defined type...[t]he second requirement is that the implication of the terms should be necessary.*[19]

Examples of terms implied by law in established contracts include the lease of a furnished house where it has long been accepted that a term will be implied that at the time of commencement of the tenancy the house will be reasonably fit for habitation.[20] The same applies in relation to a contract for the sale of land and the building of a house on it where it is implied that it will be reasonably fit for habitation.[21]

Vacant possession as an implied term

Vacant possession, when operating as an implied term, can be seen as a term implied by the courts as a matter of law. Various cases have confirmed this, including the decision in *Cook* v *Taylor*,[22] in which Simonds J said in general terms that:

> *where a contract is silent as to vacant possession, and silent as to any tenancy to which the property is subject, there is impliedly a contract [to the effect] that vacant possession will be given on completion.*[23]

---

[16] Furmston, MP *The Law of Contract* (London, Butterworths Law, 2nd revised ed, 2003) section 3.21.

[17] *Shell UK* v *Lostock Garages* [1977] 1 All ER 481, per Lord Denning MR at 487.

[18] See *Industrie Chimiche Italia Centrale and Cerealfin SA* v *Alexander G Tsavlisis & Sons Maritime Co, The Choko Star* [1990] 1 Lloyd's Rep 516, per Slade LJ at 526.

[19] See *El Awadi* v *BCCI* [1989] 1 All ER 242, per Hutchinson J at 253.

[20] See *Smith* v *Marrable* [1843] 11 M&W 5; *Wilson* v *Finch Hatton* [1877] 2 Ex D 336 and *Collins* v *Hopkins* [1923] 2 KB 617.

[21] See *Perry* v *Sharon Development Ltd* [1937] 4 All ER 390; *Lynch* v *Thorne* [1956] 1 WLR 303 and *Hancock* v *BW Brazier (Anerley) Ltd* [1966] 1 WLR 1317.

[22] *Cook* v *Taylor* [1942] Ch 349, per Simonds J at 352.

[23] ibid at 352.

In *Farrell* v *Green*[24] it was held that if the parties agree for vacant possession when a tenant is in the property, and do not mention vacant possession in the contract, it will not be ineffective for omitting a material term since vacant possession is implied into it by law. In *Midland Bank Ltd* v *Farmpride Hatcheries Ltd*[25] it was said that:

> *Prima facie a prospective vendor of property offers the property with vacant possession unless he otherwise states and that would ordinarily be implied in the contract of sale in the absence of stipulation to the contrary.*[26]

In *Edgewater Developments Co* v *Bailey*, it was noted that 'where nothing was said about possession it was often said that there was an implication that property was to be sold with vacant possession'.[27]

The implication of vacant possession is logical given that the seller and buyer relationship is embodied in a sale and purchase contract which is of an established and defined type. When an immediate right of possession is being conveyed (that is, the sale is not of an estate in reversion), it would be unthinkable for a contract for the sale of land or property not to provide the buyer with a right to possession of the estate in land on completion. In such cases, receiving vacant possession is an integral part of the contract.[28]

Impediments to vacant possession

It has been suggested that where a property is sold with vacant possession, the vendor has to satisfy the purchaser that there is no adverse claimant and no occupier of the premises at the relevant time.[29] It has been stated that:

> *An undertaking that vacant possession will be given is usually taken to mean that possession will be given free from any occupation by the vendor or a third party and free from any claim to a right to possession of the premises.*[30]

Over time, case law has tended to take the generic view that any impediment which prevents the purchaser from obtaining the quality of possession for which he or she had contracted, will constitute a breach of the seller's obligation. Indeed, case law identifies various potential obstacles to the receipt of vacant possession which can be divided into several different categories.

---

[24] *Farrell* v *Green* [1974] 232 EG 587.

[25] *Midland Bank Ltd* v *Farmpride Hatcheries Ltd* [1981] 2 EGLR 147.

[26] ibid, per Shaw LJ at 148.

[27] *Edgewater Developments Co* v *Bailey* [1974] 118 Sol Jol 312, per Lord Denning RM at 314.

[28] Williams, TC 'Sale of Land with Vacant Possession' (1928) 114 *The Law Journal* 339.

[29] *Williams on Vendor and Purchaser* (London, Sweet and Maxwell, 4th ed, 2008) p 201.

[30] See *Emmett on Title* (London, Sweet & Maxwell, 19th revised ed, 2008) para 6.006.

TANGIBLE IMPEDIMENTS

The most common example of an impediment to vacant possession is when items that should have been removed by the seller or party required to give vacant possession are left at a property on completion. There has been a plethora of case law dealing with the non-procurement of vacant possession in these terms.

Some of the very oldest case law, concerning the service of summons in ejectment cases (the removal of persons for the non-payment of rent or other contractual amounts falling due under a lease or other legal document) dealt (somewhat crudely) with vacant possession not being given due to items being left in the premises on completion. In *Savage* v *Dent*,[31] beer was left in a cellar by the party required to give vacant possession on completion and this was held to breach the obligation to give vacant possession. In *Isaacs* v *Diamond*,[32] furniture and goods remaining on the premises at completion was held to be a breach of the obligation to give vacant possession. In each case, the items being left at the premises were seen to be consistent with the seller keeping possession of the premises for his own purposes.

PERSONS IN OCCUPATION

A second common obstacle to the receipt of vacant possession is persons remaining in the property on completion. There is a wealth of case law confirming that the presence of an existing tenant or other occupier at the premises on completion will prevent vacant possession being given. This is commonly because the lease is still continuing (i.e. the party has contractual or statutory rights to remain in occupation of the property) or because other persons prevent the delivery of vacant possession on completion.[33]

In *Royal Bristol Permanent Building Society* v *Bomash*,[34] the purchaser agreed to buy two houses, vacant possession of which was to be given on completion. When the day fixed for completion arrived, the houses were occupied by someone who was 'holding over' (that is, continuing to occupy on the same terms of the tenancy which had technically come to an end by effluxion of time). It was held that the vendor was in breach of his obligation to give vacant possession on completion and damages were awarded accordingly.

---

[31] *Savage* v *Dent* [1736] 2 Stra 1064.

[32] *Isaacs* v *Diamond* [1880] WN 75.

[33] For a discussion of the problems of so-called 'sitting tenants', see Stocker, J 'The Problem of the Protected Sitting Tenant' (1988) 85 *Law Society Gazette* 14 and the Legal Update in (1988) 85 *Law Society Gazette* 36.

[34] *Royal Bristol Permanent Building Society* v *Bomash* [1886–90] All ER Rep 283 and *Engell* v *Finch* [1869] LR 4 QB 659.

Case law has historically taken an inconsistent view when the persons in occupation have no lawful claim to possession of the property (for example, trespassers),[35] but it has been established that unlawful occupiers will cause there to be a breach of the obligation to give vacant possession at the material time (in a similar manner to lawful occupiers).[36]

LEGAL IMPEDIMENTS

The third main type of impediment to the receipt of vacant possession will be an obstacle of a legal nature. Examples include the transfer of a strip of land subject to dedication as a public highway. In *Secretary of State for the Environment* v *Baylis and Bennett*,[37] it was held that vacant possession of the land could not be given because the highway authority had the right to possession rather than the owner of the sub-soil. Other instances include a property (with an existing first floor tenancy) being sold with 'vacant possession of the ground floor' but with a Housing Act notice limiting occupation of the whole house to one household, thus preventing the delivery of vacant possession as far as the ground floor was concerned.[38] Similarly, if a property remains subject to a third party lease or other tenancy that has not been properly surrendered, this will prevent the delivery of vacant possession on completion.[39] A number of other cases concern the property in question being compulsorily purchased or requisitioned in some way after the exchange of contracts, ostensibly creating a legal obstacle which prevents the giving of vacant possession on completion.

Defining the obligation

As the obligation to give vacant possession is relevant to so many transactions, and so widely used and referred to in legal agreements, it could be reasonable to assume that the concept would have a clear, settled and agreed definition. This is far from the case. Whilst of universal use, the term is one of the more misunderstood among all property terms, with few being able to explain what procuring vacant possession

---

[35] See *Sheikh* v *O'Connor* [1987] 2 EGLR 269.

[36] Megarry, W and Wade, W *The Law of Real Property* (London, Sweet and Maxwell, 7th ed, 2008) p 672 state that 'the better view is that it is the duty of the vendor to evict trespassers'. See also *Cumberland Consolidated Holdings Ltd* v *Ireland* [1946] KB 264. The issue of unlawful occupiers is discussed in more detail in chapter 6.

[37] *Secretary of State for the Environment* v *Baylis and Bennett* [2000] 80 P&CR 324.

[38] *Topfell Ltd* v *Galley Properties Ltd* [1979] 1 WLR 446.

[39] *Weir* v *Area Estates Ltd* [2009] All ER (D) 189 (Dec).

actually means or involves. Over time, lawyers, surveyors, judges and laymen have all struggled in seeking to interpret what the obligation actually refers to in cases where the procurement of vacant possession is a decisive issue.

HISTORICAL REFERENCES

The expression 'vacant possession' when used in the everyday conventional sense does not seem to be a term that has any great foundation in theoretical analysis, or historical root. Indeed, in the case of *Cumberland Consolidated Holdings Ltd* v *Ireland*,[40] counsel for the defendant summarised by stating that:

> ...*there are two classes of cases in which the question arises of what is meant by 'vacant possession'. The first class to which Savage v. Dent and Isaacs v. Diamond belong are cases relating to service of proceedings for recovery of land when personal service cannot be effected. In those cases it is essential that the premises by means of which substituted service is to be effected shall be completely deserted...As between vendor and purchaser, however, vacant possession involves the absence of any adverse claim by anyone else to a right in respect of the property being sold.*[41]

Counsel for the claimant clarified how a lack of authority existed as to the obligation:

> '*vacant possession' is not limited in meaning to the absence of any adverse claim. This limited meaning only applies to cases relating to substituted service and that is because in dealing with service the finding of a person is essential and substituted service can only be made on deserted premises. That does not assist in determining the meaning of 'vacant possession' as between vendor and purchaser, a matter not decided by any authority.*[42]

AN UNDEFINED TERM

It was correct for counsel to state that the expression 'vacant possession' had never been authoritatively defined. It has been noted by the judiciary that the law surrounding vacant possession is an area deficient in legal authority,[43] with various leading Counsel struggling, in vain, to cite relevant case law to support the legal

---

[40] *Cumberland Consolidated Holdings Ltd* v *Ireland* [1946] KB 264 at 268.

[41] ibid at 268, per Ashworth (for the defendant). Emphasis added.

[42] ibid at 268, per Heilpern (for the claimant). Emphasis added.

[43] See *Sheikh* [1987] 2 EGLR 269.

positions that they advance. Various judges have grappled to explain exactly what is meant by vacant possession. Further, the meaning of the words 'vacant possession' have been said to vary according to the context in which they are used,[44] providing little certainty to those who are eager to ensure that vacant possession is procured at the relevant time (for example, commercial tenants) but, perhaps, hope to those who would rather argue that it has not been (such as commercial landlords who want their current tenant to remain liable under the lease), as discussed above.

From the case law, it is apparent that where a vendor expressly or impliedly contracts to convey an estate in land free from incumbrances, it has been established that it is, in principle, a term of the contract that the purchaser shall on completion be entitled to actual (and not constructive) possession.[45] It has been said that:

> the phrase 'vacant possession' is no doubt generally used in order to make it clear that what is being sold is not an interest in a reversion.[46]

This would seem to imply that vacant possession is a legal issue inextricably interrelated with the passing of possession itself. In practice, however, it is clear that vacant possession has been held not to have been given if the purchaser cannot enjoy the right of possession without first having to take legal action itself.[47] Indeed, the term 'vacant possession' would seem to go beyond just a legal transfer of possession and be concerned also, on a practical level, with actual occupation (in a factual and practical sense) of the property in question:

> ...the right to actual unimpeded physical enjoyment is comprised in the right to vacant possession...[48]

As will be shown, the tests that case law has developed to determine whether vacant possession has been given are objective in nature and are concerned with whether the purchaser (or party with the right to vacant possession on completion) can occupy without difficulty or objection. The courts are required to determine whether a physical or legal impediment substantially prevents or interferes with the enjoyment of a substantial part of the property. These embody the practical dimension of vacant possession as a factual, as well as legal, matter. Chapter 5 proposes a model of vacant possession which properly accounts for the legal and factual aspects of the obligation.

---

[44] According to *Topfell* [1979] 1 EGLR 161, per Templeman J at 162. The intention of the parties will also be a materially relevant consideration.

[45] *Hughes* v *Jones* [1861] 3 De GF & J307 and *Horton* v *Kurzke* [1971] 1 WLR 769, per Goff J at 771.

[46] *Cumberland* [1946] KB 264, per Lord Greene at 270.

[47] This was a point discussed in *Sheikh* [1987] EGLR 269.

[48] *Cumberland* [1946] KB 264, per Lord Greene at 272.

Outline of chapters

It is quite apparent that the obligation to give vacant possession affects the majority of land transactions in England and Wales, and is of relevance to a wide range of people and other legal entities. Given the lack of understanding associated with the term, it is not surprising that the nature, scope and extent of this commonly misunderstood obligation to give vacant possession causes a wide range of problems to the thousands of people that the term affects each day in everyday property transactions. This book addresses such issues and provides clarity and explanation on the various issues to which vacant possession is relevant.

CHAPTER 2

Chapter 2 outlines in more detail the problems with the vacant possession construct in the current legal system. This is in order to show an appreciation of the magnitude of the issues associated with the term. The chapter explains how problems arise with seeking to interpret the obligation at every stage in terms of whether the obligation has arisen, what the obligation relates to, what can constitute a breach of the obligation, and whether there has, in fact, been a breach. Even then, the adequacy of the remedies available to the parties is shown to be questionable. In doing so, this chapter sets the issues in context before these are subsequently analysed, and explained, in more detail in later chapters.

CHAPTER 3

In chapter 3 the role of vacant possession as a term of a contract for the sale or lease of land is explored by reviewing its interaction with other terms (or conditions). In particular, the interaction between a special condition providing for vacant possession and two other more standard clauses (namely, 'subject to local authority requirement' clauses and 'no annulment, no compensation' clauses) is discussed through an analysis of relevant case law. The decisions in case law on such interactions were inconsistent prior to 1979 (when the landmark decision in *Topfell Ltd* v *Galley Properties Ltd*[49] was laid down). This decision established the precedence of a special condition for vacant possession over other competing contractual conditions. The chapter highlights, however, how a lack of authority continues to leave the position unclear in cases where the vacant possession condition is only a general condition, or is merely implied into the contract, a matter often not appreciated by legal draftsmen.

---

[49] *Topfell* [1979] 1 WLR 446.

Chapter 4

Having established that vacant possession is normally a term in a contract, chapter 4 explains the place of vacant possession (as a term) in 'standard conditions of sale' which are routinely incorporated into the majority of contracts for the sale and purchase of freehold land (and leasehold estates and interests) by reference. This is the most common context in which the term 'vacant possession' will be encountered by lawyers and other professionals.

The incorporation of a term for vacant possession as either a general or special condition is shown to be highly relevant to the precedence (or otherwise) of the obligation with respect to other terms of the contract, and the consequent obligations and liabilities of the parties involved. The chapter also highlights a number of specific issues relevant to practitioners when incorporating the current editions of the conditions of sale into standard sale and purchase contracts, given the differing wording of the special conditions for vacant possession in each of the respective editions of the conditions of sale (namely, residential and commercial) at this time.

Chapter 5

In chapter 5, an understanding of the obligation to give vacant possession as comprising both a legal and factual dimension is explained. Vacant possession, which necessarily concerns actual (*de facto*) possession on completion, pursuant to the legal right to possession which is transferred with the estate in land (*de jure*), is contrasted with notions of constructive possession (i.e. possession *otherwise* than by actual occupation) in order to demonstrate why the legal and factual elements are intrinsic to the obligation. This chapter therefore expounds the content of the obligation as a matter of fact and law, thus aiding understanding of its operation in a practical context.

Chapter 6

In chapter 6, an understanding of the legal and factual dimensions of the obligation is then applied to the three main categories of obstacle to the receipt of vacant possession (namely, tangible impediments, persons in occupation and legal impediments), and the extent to which a breach of the obligation can occur. For each of the categories of impediment discussed, relevant case law is cited and evaluated. The respective limbs of the tests are explained, along with the legal and factual considerations that must be taken into account in determining whether a breach of the obligation to give vacant possession has arisen in any given case.

CHAPTER 7

Having discussed the tests to determine a breach of the obligation, chapter 7 identifies which potential obstacles to the receipt of vacant possession may be outside the scope of the obligation. For example, traditionally, fixtures (unlike chattels) are not seen to be relevant to the vacant possession obligation. This chapter demonstrates that whilst the nature of the obligation can be broadly explained, the obligation's scope and extent remains unclear in a variety of respects. The analysis suggests that traditional distinctions between, for example, fixtures and chattels, are not relevant in the context of vacant possession, and that the crucial issue relates to whether the item is a substantial obstacle to the receipt of 'possession'. As such, the state and condition of a given premises is shown to be potentially relevant to the vacant possession obligation (as are any applicable contractual conditions which may modify, or negate, such an obligation). By contrast, so-called 'non-possessory' or 'lesser interests' (such as certain *profits*) are explained as not being relevant to the vacant possession determination, given that (by their very nature) they do not amount to impediments that can affect the receipt of 'possession'.

CHAPTER 8

Chapter 8 starts by addressing practical timing issues concerning the point of completion itself, and how this should tie in with the obligation to give vacant possession. If vacant possession is not given at the moment of completion, then technically a breach will have occurred.

When a breach of the obligation to give vacant possession is held to have taken place, the next determination for a court will be the remedy or relief that can be awarded to the successful party. The remedy normally awarded to an injured party for a breach of the obligation to give vacant possession will be damages, which can often be largely unsatisfactory to a purchaser who, having already paid the purchase money before finding that the property is not vacant, will be unable to occupy the property as he or she wished to. The chapter explores the remedies currently available and proposes both a contractual definition of vacant possession, and enhanced damages provisions, which can assist parties when a breach of the obligation may have occurred.

CHAPTER 9

Chapter 9 focuses on issues of vacant possession that are specific to the leasehold context and, in particular, the compliance with break options in leases which are conditional on the giving of vacant possession. The issues for lawyers and surveyors when seeking to ensure that a break can be successfully operated, along with the consequences for both landlord and tenant, are discussed in detail. It is also explained how the obligation to give vacant possession can form part of a more general requirement on the tenant to 'yield-up' the premises upon exercise of the break right. The chapter also considers other leasehold issues to which vacant possession is relevant, including the assumption of vacant possession on rent review and corresponding implications for valuation.

Chapter 10

To consolidate understanding of the issues of vacant possession which have been explained throughout the preceding chapters, chapter 10 is made up of three case studies, with suggested answers and points to note, that address matters relating to:

1. Vacant possession in the freehold context.
2. Vacant possession in the leasehold context.
3. Vacant possession and the state and condition of the property.

Chapter 2

Issues in Vacant Possession

As alluded to in chapter 1, there is no generally accepted understanding of the nature, scope and extent of the obligation to give vacant possession. Not surprisingly, this causes a wide range of issues for the cross-section of people who are affected by the obligation. In order for all the problematic issues currently associated with the term to be fully appreciated, it is appropriate to provide an insight into the real life scenarios in which the expression 'vacant possession' is used and applied. This will assist in explicating the difficulties faced by this wide variety of people on a day-to-day basis, and the risks and responsibilities which overshadow the giving of vacant possession.

19

Whilst the nature, scope and extent of the obligation, and the remedies that flow from a breach of the term, can be properly described as opaque (as shall be discussed shortly), these only become relevant if an obligation to give vacant possession has actually arisen — something itself which cannot be assumed to be lucid or self evident in every case. Starting with consideration of whether the obligation has even arisen, this chapter seeks to set the various issues relating to vacant possession in context, before these are subsequently analysed in more detail in later chapters.

Has the obligation arisen?

As previously noted, the issue of vacant possession can arise on the sale of freehold land (for example, the transfer of estates in fee simple) and upon the grant, transfer and termination of leases and other tenancies. The term is also used in respect of informal agreements to occupy by consent (for example licences), even though use of the term in such a context is arguably inappropriate (as explained in chapter 1). With respect to leases, procuring vacant possession is generally seen as an essential, and likely, element of the obligation to yield-up the premises at the contractual termination of the lease. Further, vacant possession itself can also be an express precondition for the exercise of a tenant's break option in a lease or tenancy.[1] From the outset, it is important to be aware of whether an obligation to give vacant possession has arisen and, if so, on what basis. This is because how the obligation arises directly determines what that obligation will encompass.

Vacant Possession Clauses

It is common for a contract for the sale and purchase of land to include an express obligation to give vacant possession on completion, for example:

> *the property is sold with vacant possession on completion.*

Often, however, the contract will simply prescribe that the property must be 'transferred free from all incumbrances'. It is unclear as to whether this can be construed as simply referring to legal incumbrances on title or whether this may actually relate to providing vacant possession by referring to any and all impediments that could affect enjoyment of the property on completion. The meaning of the term 'incumbrances' has been said to vary according to the circumstances.[2]

---

[1] See Martin, J 'Tenant's Break Options' (2003) 153 *New Law Journal* 759 where the requirement to give vacant possession when operating a break option in a lease is discussed.

[2] See *Belvedere Court Management Ltd* v *Frogmore Development Ltd* [1997] QB 858 at 887.

Whilst, in practice, the question is ultimately one of the intentions of the parties as shown by the contract,[3] it is sometimes difficult to determine the intention of the parties with reference to any given clause of this kind. This is especially the case when a given clause is read in conjunction with the standard conditions of sale.

As discussed in chapter 4, over time many editions and versions of standard conditions of sale have been incorporated into sale and purchase contracts only by reference, with the various generic terms therefore not having been specifically considered by the contracting parties. As such, even when the contract clearly includes an express obligation to give vacant possession, it can be unclear as to whether such an obligation can be contradicted by these other general conditions of sale which can provide, for example, that a purchaser buys subject to notices and to anything which would have been revealed by local searches and enquiries. In *Topfell Ltd v Galley Properties Ltd*,[4] it was held that an express obligation to give vacant possession (appearing as a special condition in a sale and purchase contract) could not be contradicted by the usual general conditions of sale. This decision is, however, in direct conflict with other decisions in case law which subordinated an express term for vacant possession in favour of other conditions.[5] Over time there has been an apparent difficulty in interpreting the extent of an obligation to give vacant possession when such an obligation has been found to interact with (or be qualified by) other conditions of sale. Chapter 3 explains the position in more detail and how the precedence of the obligation has now been established in most, but not all, cases.

Even if other contractual conditions can (in principle) modify an obligation to give vacant possession, it is unclear as to which will (and will not) be relevant to the obligation. Indeed, it has been argued that standard conditions relating to the state and condition of the property can operate to affect or otherwise modify an obligation to give vacant possession,[6] but there remains no actual authority on the position where the seller's inability to give vacant possession is due to the physical state of the property.[7] Chapter 7 explains the scope and extent of the obligation in more detail in this regard.

IMPLIED OBLIGATIONS

Whilst a vacant possession obligation can appear as an express clause in the contract, it is common for conditions to fail to cater for vacant possession in such terms.

---

[3] See *Lake v Dean* [1860] 28 Beav 607 and *Re Crosby's Contract* [1949] 1 All ER 830.

[4] *Topfell Ltd v Galley Properties Ltd* [1979] 1 EGLR 161.

[5] See, for example, *Curtis v French* [1929] 1 Ch 253 and *Korogluyan v Matheou* [1975] 30 P&CR 309.

[6] See *Topfell Ltd v Galley Properties Ltd* [1979] 1 WLR 446.

[7] According to Harpum, C 'Vacant Possession — Chamaeleon or Chimaera?' (1988) *Conveyancer and Property Lawyer* 324, 400 (CH) *cf Hynes v Vaughan* [1985] 50 P&CR 444, per Scott J (and discussed in detail in chapter 7).

Ordinarily, this will mean that vacant possession will be no more than an *implied* term of the contract.

Cheshire notes that 'it is an implicit term of the contract of sale that vacant possession shall be given to the purchaser on completion'.[8] In *Cook* v *Taylor* it was held that where a contract is silent as to vacant possession, and silent as to any tenancy to which the property is subject, there is impliedly a contract that vacant possession will be given on completion.[9] The case concerned a vendor who had entered into a written agreement (which incorporated the Law Society's Conditions of Sale 1934) with a purchaser for the sale of a freehold property. The agreement contained no reference to vacant possession, but particulars containing the statement 'vacant possession on completion' had been delivered by the vendor's agents to the purchaser. Before the date fixed for completion, the house was requisitioned by the Government under the Defence (General) Regulations 1939 (which were war time provisions for the compulsory acquisition of property). The judge decided that the particulars were used in connection with the contract and were incorporated therein by the Law Society's Conditions of Sale; therefore there was a contract expressly to sell the property with vacant possession. Apart from that, however, it was held that according to the general law there was an implication that the property was to be sold with vacant possession, that is, the law will *imply* an obligation to give vacant possession in the absence of an express clause providing for vacant possession.

The decision in *Cook* was later followed in *Re Crosby's Contract, Crosby* v *Houghton*[10] and other case law. For example, in *Midland Bank Ltd* v *Farmpride Hatcheries Ltd*,[11] it was held that 'prima facie a prospective vendor of property offers the property with vacant possession unless he otherwise states and that would ordinarily be implied in the contract of sale in the absence of stipulation to the contrary'. Lord Denning in *Edgewater Developments Co* v *Bailey* said that 'where nothing was said about possession it was often said that there was an implication that property was to be sold with vacant possession'.[12] As such, vacant possession being an implied term of a contract has been established in various judgments of the court.

Irremovable impediments

When an obligation to give vacant possession has arisen impliedly, it is important to note that the implied assumption will be subject to specific circumstances and actual

---

[8] Cheshire, GC and Burn, EH *Modern Law of Real Property* (London, Butterworths, 11th ed, 1976) p740. See also Farrand, JT *Conveyancing Contracts: Conditions of Sale and Title* (London, Oyez Publications, 1964) pp 209–213.

[9] *Cook* v *Taylor* [1942] Ch 349, per Simons J at 352.

[10] *Re Crosby's Contract* [1949] 1 All ER 830.

[11] *Midland Bank Ltd* v *Farmpride Hatcheries Ltd* [1981] 2 EGLR 147, per Shaw LJ at 148.

[12] *Edgewater Developments Co* v *Bailey* [1974] 118 Sol Jol 312.

knowledge of the parties. For example, where one party is aware, when entering into a contract, that the interest is subject to some impediment to vacant possession, case law suggests that if the purchaser knows that the obstacle to the receipt of vacant possession is irremovable, then the implied obligation to give vacant possession will not extend so as to include that obstacle (unless expressly agreed otherwise).

In *Timmins* v *Moreland Street Property Co Ltd*,[13] the vendor was bringing an action for damages for repudiation of contract to buy certain freehold property under an oral agreement. The defendants contended, *inter alia*, that the documents relied on a memorandum that did not sufficiently comply with section 40 of the Law of Property Act 1925 and (amongst other matters) omitted to state that the property was sold subject to a lease. It was held that the omission to refer to the lease did not vitiate the memorandum and that the defendants, by virtue of their knowledge (that the plaintiff's interest was subject to the lease) when they entered into the contract, were precluded by implication of law from objecting to take the property subject to the lease (regardless of whether it was referred to in the memorandum). Their knowledge of the lease (as an irremovable obstacle) on exchange meant that an implied obligation to give vacant possession did not extend so as to include that obstacle on completion and, as such, the lease did not constitute a breach of the obligation to give vacant possession when it was in place at completion.[14]

Unhelpfully, there is no actual guidance as to what is deemed 'irremovable' in this context. In *Hughes* v *Jones*,[15] a lease was said to be irremovable but a seller may agree to surrender the interest between exchange and completion and, in that regard, a purchaser could legitimately claim to have assumed that to be removable before completion, causing the implied obligation to encompass the leasehold interest. Further, in *District Bank Ltd* v *Webb*,[16] a lease was not treated as an incumbrance. Whilst this case was in the context of defects of title, the principles relevant to vacant possession closely follow those applicable to a vendor's duty to disclose latent defects in title, which can be implied in similar terms.[17] It can therefore be argued that a lease or other tenancy or agreement pertaining to occupation may not *always* be an irremovable incumbrance, even though the default position is likely to be that it will be.[18]

---

[13] *Timmins* v *Moreland Street Property Co Ltd* [1958] Ch 110.

[14] *cf Farrell* v *Green* [1974] 232 EG 578 at 589 which was decided *per incuriam* on this point.

[15] *Hughes* v *Jones* [1861] 3 De GF & T 307. See also *Re Englefield Holdings Ltd* v *Sinclair's Contract* [1962] 1 WLR 1119. This is obviously subject to contra-indications or other intentions of the parties as shown by the contract.

[16] *District Bank Ltd* v *Webb* [1958] 1 WLR 148.

[17] See Harpum, 'Vacant Possession — Chamaeleon or Chimaera?' (1988) *Conveyancer and Property Lawyer* 324, 400.

[18] See also *Caballero* v *Henty* [1874] LR 9 Ch 447.

In terms of what will amount to 'knowledge' of an irremovable obstruction for these purposes, it is difficult to determine whether a party will be deemed to have sufficient knowledge of the obstruction. In *Lake* v *Dean*,[19] notice that the property was 'in the occupation of a third party' (as opposed to the disclosure of the express terms of the given tenancy) was *not* considered to constitute disclosure of a tenancy that was sufficient to amount to knowledge of an irremovable obstruction which could modify an implied obligation to give vacant possession.[20] Constructive notice of the existence of a tenancy may, however, be imputed to the buyer. For example, if the purchaser knows that a person is in occupation of the property, the purchaser is presumed to know the rights of the occupier, and where the occupier has such a legal tenancy, the buyer, according to *Hunt* v *Luck*,[21] will take subject to that tenancy. The question of whether a purchaser has sufficient knowledge of an irremovable impediment, which the property will be transferred subject to, is therefore not always a straightforward determination, and a purchaser is wise to ensure that he or she is fully aware of any potential impediment prior to exchange. It would be dangerous for a purchaser to seek to rely on limited knowledge of a third party interest to claim that he or she was not actually aware of the impediment, even though the extent of one's knowledge will be determinative in any given case, which will always turn on the precise facts in issue.

If a purchaser does not know (or is not deemed to know) of an irremovable obstruction at the point of exchange, then the implied obligation *will* encompass such an obstacle. If that obstacle remains in place on completion (which is most likely, if it is irremovable), then it will constitute a breach of the obligation to give vacant possession.

Removable impediments

If, at the time of contract, the purchaser knew only of a *removable* obstacle, then the implied obligation to give vacant possession will *not* be deemed to exclude such an obstacle (i.e. the implied obligation *will* encompass such an obstacle). If the removable obstacle is still on the premises on completion, the obligation to procure vacant possession will have been breached.

In *Norwich Union Life Insurance Society* v *Preston*,[22] a mortgagor was ordered, in July 1956, to deliver to the mortgagees' possession of the mortgaged premises within 28 days. In September 1956, the order was served on the mortgagor, and in December 1956, possession not having been given, a writ of possession was issued. In January 1957, the sheriff's officer evicted the mortgagor. The mortgagor left his furniture in the mortgaged

---

[19] *Lake* [1860] 28 Beav 607.

[20] This decision pre-dates the Landlord and Tenant Act 1954 and other relevant legislation relating to the secured rights of occupiers; it remains, however, an illustrative example of the issues that can, in principle, arise in this regard.

[21] *Hunt* v *Luck* [1902] LRA 2002.

[22] *Norwich Union Life Insurance Society* v *Preston* [1957] 1 WLR 813.

premises and refused to remove it, contending that the order for possession was spent. The mortgagee applied for an order that the mortgagor should remove his furniture within four days. It was held that the mortgagee was entitled to this order because the mortgagor had not given vacant possession in compliance with the order. Leaving his furniture on the premises was seen by the Court as to amount to the mortgagor claiming a right to continue to use the premises for his own purposes. It did not matter that the mortgagee was *aware* of the presence of the furniture at the material time, for such items were clearly capable of being removed. The implied obligation to give vacant possession *did* therefore encompass the removable obstruction, which was known about on exchange, and as the obstruction was not removed there was accordingly a breach by the mortgagor on completion. As such, this case confirms that knowledge of *removable* obstructions will be irrelevant when interpreting the scope of an implied obligation to give vacant possession, which will encompass removable obstacles.

EXPRESS OBLIGATIONS

The position on removable and irremovable obstructions, with respect to the implied obligation to give vacant possession, can be contrasted with the position where there is an express obligation to give vacant possession. Here, the position is entirely different. It has been held that an express obligation to give vacant possession *will* prevail regardless of the nature of any known potential impediment to vacant possession (removable or irremovable).

In *Sharneyford Supplies Ltd* v *Edge*,[23] the plaintiff purchased land from the defendant by a contract which incorporated the Law Society's General Conditions of Sale (1973 revision). It provided by a general condition that, unless the special conditions otherwise provided, the property was sold with vacant possession on completion. The plaintiff, aware that the land was occupied, had stressed from the outset that vacant possession was required and had received answers to pre-contract enquiries from the defendant that the occupants had no right to remain in possession. The occupants refused to vacate the land on completion. The express obligation to give vacant possession meant that the defendant was in breach, even though the plaintiff purchaser knew, at the time the contract was formed, of an irremovable obstruction to the delivery of vacant possession (the lease).[24] As Parker LJ observed:

> If ... a vendor sells land, which he knows is subject to a tenancy, and contracts specifically to sell with vacant possession, he makes in effect, a specific promise that he will get the tenants out.[25]

---

[23] *Sharneyford Supplies Ltd* v *Edge* [1987] Ch 305.

[24] See also *Hissett* v *Reading Roofing Co Ltd* [1996] 1 WLR 1757.

[25] *Sharneyford* [1987] Ch 305, per Parker LJ at 325.

An express vacant possession clause makes knowledge of potential obstructions (whether removable or irremovable) irrelevant.

The table below summarises the position in this regard.

Table 2.1 Express/Implied Vacant Possession Conditions and Forms of Obstacle

	Removable obstacles	Irremovable obstacles
Express obligation to give vacant possession...	Will cover these obstacles, whether known of on exchange or not.	Will cover these obstacles, whether known of on exchange or not.
Implied obligation to give vacant possession...	Will cover these obstacles, whether known of on exchange or not.	If known (or deemed to be known) of on exchange: Will *not* cover these obstacles. If not known (or not deemed to be known) of on exchange: Will cover these obstacles.

PRACTICAL IMPLICATIONS

From a practitioner's point of view, ascertaining *how* the obligation to give vacant possession has arisen (by express provision or implied by law) is therefore essential in order to determine what the obligation can be deemed to encompass.

An example illustrates the issues. Imagine that a seller contracts to convey land to a purchaser and that, upon exchange, the purchaser knew of large items in the property (which we will assume were clearly *irremovable*). The only possible reference to vacant possession in the contract is a clause stating that:

the property is sold free from legal incumbrances on completion.

It is therefore unclear as to whether this refers to vacant possession expressly. On completion, large items remain on the property and the purchaser claims that the seller has not given vacant possession because of the existing presence of these items. The purchaser claims that the seller expressly contracted to provide vacant possession, and is therefore in breach. Conversely, the seller claims that the clause in question referred only to title issues and therefore that any obligation to procure vacant possession is implied. The seller further claims that because the purchaser was aware of the large items in the property on exchange, and that they were *irremovable*, the implied obligation to give vacant possession did not extend to include these items. The seller therefore contends that it is not in breach. How the obligation arose (expressly or impliedly) will determine whether the seller or purchaser is correct in his or her respective claims as to whether the alleged obstruction constitutes a breach of the obligation to give vacant possession.

This scenario is often encountered and it is common for contracts to include other ambiguous clauses that could be construed as referring expressly to the procurement of vacant possession, for example:

- *'The purchaser will be entitled to actual possession on completion'.*
- *'On completion the purchaser is to have undisturbed enjoyment of the property'.*
- *'The seller will convey the land free from all legal impediments'.*
- *'The purchaser will be entitled to free occupation on completion'.*
- *'The legal right to unencumbered enjoyment...will pass on completion'.*

It is not clear as to whether any of these would amount to an express obligation to give vacant possession, and the context of the clause in the contract and the intention of the parties would have to be considered in making any determination.

IMPLIED OBLIGATION IN LEASES

In the leasehold context, a tenant may have the right to exercise a break option in a lease and bring that lease to an end earlier than the contractual termination date. Practically speaking, it is advantageous that a tenant avoids the procurement of vacant possession as a pre-condition for exercise of the break option (and that any agreement to provide vacant possession is entirely separate from the exercise of the break option). This way, the tenant will not be prevented from exercising the break if vacant possession cannot be given and the landlord, rather than seeking to contend that the lease is still continuing, will instead be forced to rely on other remedies to deal with the vacant possession issue when the lease has terminated.

In practice however, vacant possession is often an express pre-condition for the exercise of a break option, or alternatively, can be implied by virtue of a more general pre-condition upon exercise of the break option, namely that the tenant has 'materially complied' with all tenant covenants under the lease. Where a break option is conditional upon the tenant's material compliance with covenants up to and *including* the break date, this will almost certainly encompass the 'yielding-up' obligation that will take effect on termination of the lease. It is likely that the yielding-up obligation will require the return of the premises with vacant possession of some description.[26]

For example, imagine that a tenant has trouble paying the rent on a lease and decides to move to smaller premises. The tenant exercises a break option in its lease with the landlord which is conditional upon compliance with all covenants in the lease up to and including the break date. On the break date the premises are not

---

[26] See chapter 9 for a discussion of the scope of the requirement to give vacant possession in the context of yielding-up under a lease.

cleared of tenant's chattels. The tenant claims that vacant possession was not a condition of the break option in the lease. The landlord takes the position that the condition that the tenant must have complied with all covenants up to and including the break date includes the tenant's covenant to yield-up at the end of the lease, which itself includes the return of the premises with vacant possession. If the landlord is correct, then the tenant will remain liable under the terms of the lease, which continues.

In this regard, and as highlighted in chapter 1, a landlord can use the issue of vacant possession to prevent the tenant exercising a contractual break option in a lease if the landlord would prefer the lease to continue. In such a case, it is likely that a tenant will not appreciate this until the issue is raised, normally *after* the break date, when the tenant's opportunity to give vacant possession will have passed.

Is the Obligation Engaged?

It is clear that, in everyday transactions, confusion is rife as to *when* an obligation to give vacant possession is engaged or operative and, in turn, what that obligation will actually refer to or encompass.

In the freehold context, lawyers struggle to advise clients as to whether a contractual provision amounts to an express obligation to give vacant possession and therefore, in turn, whether a purchaser's knowledge of an irremovable obstruction at the time that contracts were exchanged will be relevant to the vacant possession to which the purchaser is entitled on completion. Even then, the status of an obstruction as *removable* or *irremovable* has currently not been clarified with sufficient certainty. The potential for an obligation to give vacant possession to interact with other contractual terms, modifying or negating the term for vacant possession, is also a factor that must be evaluated in the context of the overall contract. With respect to leases, it is generally not appreciated how vacant possession can creep into the operation of break options in leases by virtue of the drafting of the lease more generally, thus causing the obligation to have arisen (almost inadvertently).

As such, the parties to a given transaction can find themselves in difficult positions given the ambiguity that can be manifest in cases of this kind. It is common to fail to appreciate the risk involved in entering into a given transaction involving vacant possession, and what responsibilities may be engaged with respect to the other party to the transaction. Chapter 3 explores how vacant possession, as a term of contract, can interact with other contractual terms and how the obligation should be interpreted with respect to other competing contractual conditions. Chapter 9 focuses on the leasehold issues which are relevant to the obligation to give vacant possession.

▌ To what does the obligation refer?

As noted in chapter 1, obstacles to the receipt of vacant possession can be divided into several different categories, each category raising issues that the law has currently failed to adequately address. It is worth considering the common types of obstacle to the receipt of vacant possession, and explaining the problems associated with each.

TANGIBLE IMPEDIMENTS

The most common example of an impediment to vacant possession is when items that should have been removed by the seller or party required to give vacant possession, are left at a property on completion. Beer in the cellar,[27] furniture and goods remaining on the premises[28] and other chattels of the party required to give vacant possession[29] have been held to breach the obligation. In each case, the items being left at the premises were seen to be consistent with the seller keeping possession of the premises for his or her own purposes.

Whilst determined by the facts of each case, historically case law never sought to set out any general principles which could be applied universally across the board, in order to determine what giving vacant possession actually means. In some cases, rubbish and leftover chattels caused a breach of the obligation, whereas in other cases items such as furniture and general items left behind were seen as irrelevant.[30] Historically, it was unclear as to whether a *de minimis* threshold operated with respect to left over items which may have explained the differing decisions reached by respective judges on ostensibly similar questions of fact.

More recently, case law has sought to develop tests to avoid ad hoc value judgments being made as to whether leftover items have prevented the procurement of vacant possession in any given particular instance. As discussed in the section entitled 'A Breach of the Obligation?' below, it is arguable whether such tests have resolved a number of issues posed by the vacant possession obligation.

Indeed, whilst more recent case law has indicated that the vacant possession obligation will be subject to a *de minimis* rule,[31] how that operates in practice and to what such a threshold refers remains unclear. The quantity of items left, their size, movability and degree and purpose of annexation would seem to be relevant factors in determining whether the items left cause a breach of the obligation to give vacant

---

[27] *Savage* v *Dent* [1736] 2 Stra 1064.

[28] *Isaacs* v *Diamond* [1880] WN 75.

[29] *Cumberland Holdings Ltd* v *Ireland* [1946] KB 264.

[30] More recently, see *Scotland* v *Solomon* [2002] EWHC 1886 (Ch).

[31] Following *Cumberland* [1946] KB 264 where the obligation was stated as being subject to such a rule.

possession. Further, case law also seems to suggest that it may be relevant to consider the location of items in, around or outside the property concerned.[32]

Another problem commonly experienced is with regard to what status items left at the property on completion may have. Disputes can arise as to whether items left behind at a property are fixtures (and therefore part of the land) or chattels (personal property of the tenant obliged to procure vacant possession).[33] Traditionally, the law has been clear that if the seller's failure to give vacant possession is due to the presence on the property of *chattels*, which affect usability of the premises, then a breach of the obligation to give vacant possession will arise if the impediment substantially interferes with enjoyment of a substantial part of the premises.[34] Conversely, fixtures (which pass as part of the land conveyed) have traditionally not been seen to be relevant to the obligation to give vacant possession, which is why the distinction has been seen as so important.[35]

A practical example illustrates such issues in context.

Imagine that a seller contracts to convey a property to a purchaser, the contract providing expressly that vacant possession is to be given on completion. The purchaser intends to grant a lease to a business tenant of the property on the same day. The proposed tenant requires occupation that day given the nature of its business. On the morning of completion the transaction completes and the purchaser is given the keys. Later in the day the purchaser meets his proposed new tenant at the premises to sign the lease and hand the keys over. Upon inspection of the property, however, the purchaser and the tenant see that furniture and other items have been left by the seller. The tenant refuses to sign the lease because the tenant says that it cannot immediately occupy the property as it needs to. Instead, it takes a lease of an adjacent unit the following week. In two months time the purchaser manages to lease out the property to a third party tenant at a rent lower than had been agreed with the proposed tenant due to a decline in the market. The purchaser claims that the seller was in breach of his express contractual obligation to give vacant possession and claims that loss has been suffered as a consequence. The seller claims that the items left were fixtures (and therefore part of the land) or, in the alternative, if they were chattels that they came within the *de minimis* threshold and therefore that no breach had arisen. The proper determination of the status of the items can be seen as a preliminary issue in seeking to establish whether the items had been left behind by the tenant unlawfully, and therefore caused a breach of the vacant possession

---

[32] *Hynes* v *Vaughan* [1985] 50 & CR 444.

[33] *Elitestone Ltd* v *Morris* [1997] 2 All ER 513, which sought to differentiate between fixtures and chattels based on their respective degree and purpose of annexation.

[34] See *Cumberland* [1946] KB 264.

[35] In chapter 7, the scope and extent of the obligation is argued to encompass more than just chattels, suggesting that the distinction is not as fundamental to issues pertaining to vacant possession.

obligation. Whether the items could be argued to have been *de minimis* would also be a relevant point to have been decided.

In the section entitled 'A Breach of the Obligation?', it is explained how, in any particular instance, actually determining what will constitute a sufficient impediment to prevent the delivery of vacant possession remains a difficult task, even following the tests that case law has sought to develop to assist in this regard.

PERSONS IN OCCUPATION

There is a wealth of case law confirming that the presence of an existing tenant or other legal occupier at the premises on completion will prevent vacant possession being given.[36] This may be because the lease is still continuing or because the party has contractual or statutory rights to remain in occupation of the property.

In *Sharneyford*[37] (referred to above), the plaintiff purchased land from the defendant under a contract that expressly provided for vacant possession on completion. The occupants refused to vacate the land and claimed the benefit of a business tenancy within Part II of the Landlord and Tenant Act 1954. The defendant was liable for not giving vacant possession at the material time.

Similarly in *Beard* v *Porter*,[38] the vendor had agreed to sell to the purchaser a dwelling-house which was occupied by a sitting tenant. The vendor expressly agreed that the purchaser was to be given vacant possession on completion. The purchase was completed, but the tenant refused to quit the house. The purchaser sued and was awarded damages for breach of the vendor's undertaking to give vacant possession on that date.

Crucially, these decisions dealt with purportedly 'lawful' claims to remain in occupation of the property (i.e. because the tenant had a statutory continuation tenancy or common law right to 'hold over').

While the traditional definitions of vacant possession refer to delivery of the property 'free from any claim to a *right* to possession of the property', it has been questioned in case law as to whether this position remains the case with respect to persons who may be in occupation with no *lawful* claim or right (for example, squatters or trespassers). There is conflicting *obiter dicta* with regard to whether people in *unlawful* occupation breach the obligation to provide vacant possession.

Some statements[39] suggests that the obligation would be breached in this situation, presumably on the basis that it is the duty of the seller (as the person responsible for providing vacant possession) to ensure that trespassers are

---

[36] *Beard* v *Porter* [1948] 1 KB 321.

[37] *Sharneyford* [1987] Ch 305.

[38] *Beard* v *Porter* [1948] 1 KB 321.

[39] Obiter comments in *Cumberland* [1946] KB 264.

evicted. In *Cumberland*[40] it was noted that a seller's duty extends to removing unlawful occupants on completion. The main issues in the case related to leftover goods at the premises but the judge considered (obiter) that the existence of a physical impediment, which substantially prevented or interfered with the enjoyment of the right of possession of a substantial part of the property, stood in the same position as an impediment caused by the presence of a trespasser.[41]

Other obiter dicta, however, suggests that a seller would not be in breach by virtue of there being a person in unlawful occupation of the property at completion. Here, receiving a property free of unlawful occupants on completion seems to be treated as a right which (in the absence of any competing legal claim) passes to the purchaser on completion (and which the purchaser can take county court action of their own accord to enforce if necessary), rather than an obligation of the seller. In *Sheikh* v *O'Connor*,[42] the vendor contracted to sell a property to the plaintiff. Most of the property was tenanted but the vendor expressly contracted to sell one of the rooms with vacant possession. After completion, the purchaser complained that the room which should have been vacant was in fact occupied by one of the tenants as a trespasser. The purchaser sued the vendor for damages for his failure to give vacant possession. One of the issues was factual and related to whether the tenant had taken possession of the room before, or after, the completion date. Deputy Judge Wheeler concluded that it had been after completion, giving judgment for the defendant. However, the judge went on to consider the position in the event that his finding of fact was incorrect (and as such the comments are obiter).

The judge accepted that a vendor who had contracted to give vacant possession did not fulfil his contractual obligation if, at the date fixed for completion, there was a third party who had a *legal* claim to possession, but he did not consider this to be the case in relation to a trespasser. In such a case he considered that it was for the purchaser to seek his remedy in the county court against the trespasser, given that the right to possession had passed to him/her on completion.[43]

Whilst current case law suggests that trespassers will be treated in similar terms to persons with lawful claims to possession,[44] earlier case law has been far from consistent on this point. Indeed, in the decision in *Sheikh* where this issue was specifically addressed, it was noted that neither counsel could find any authority which pertinently dealt with the matter before the learned judge.[45]

---

[40] ibid at 246.

[41] Supporting *Royal Bristol Permanent Building Society* v *Bomash* [1886-90] All ER Rep 283; and *Engell* v *Finch* [1869] LR 4 QB 659.

[42] *Sheikh* v *O'Connor* [1987] 2 EGLR 269.

[43] Discussed in more detail in chapter 6.

[44] *Cumberland* [1946] KB 264.

[45] *Sheikh* [1987] EGLR 269 at 271.

The possible distinction between lawful and unlawful occupiers is moreover a problem because of the way contracts and, in particular, (ostensibly) express obligations to give vacant possession, are often drafted.

For example, imagine a seller contracts to convey land to a purchaser. The only possible reference to vacant possession in the contract is a clause stating that:

the property is sold free from legal incumbrances on completion.

It is unclear as to whether this refers to vacant possession expressly. On the evening before completion, a group of new age travellers gain entry to the premises and commence using the premises as a place of refuge. The next day the seller and purchaser complete. The purchaser then goes round to the premises and realises that the unlawful persons are in occupation. The seller will claim that the clause in the contract was an express contractual undertaking to give vacant possession but, as it was restricted to 'legal' incumbrances, and the trespassers are clearly in the premises unlawfully, they do not cause the obligation to have been breached. The purchaser will claim that the seller has not given vacant possession because of the trespassers. The purchaser's position will be that, if the clause does not amount to an express contractual obligation to give vacant possession, the implied obligation to give vacant possession *will* encompass the trespassers because they were not there on exchange (i.e. the purchaser did not know of an (irremovable) obstacle to the receipt of vacant possession at the time of contract) and that the implied obligation to give vacant possession refers to the delivery of the property free from all claims (lawful and unlawful) as to a right to possession by others. It is a matter of construction of the clause as to whether the seller or the purchaser is correct.

As discussed in chapter 5, the transfer of the legal right to possession and the ability to actually occupy pursuant to that right are both part of the obligation to give vacant possession, and an understanding of these helps to explain some of the contradictory decisions in historic case law.

LEGAL OBSTACLES

It is possible that a legal obstacle may prevent the delivery of vacant possession on completion. As referred to in chapter 1, examples include the transfer of a strip of land subject to dedication as a public highway.[46] This is on the basis that the highway authority has the right to possession, rather than the owner of the sub-soil. Other instances would include a property (with an existing first floor tenancy) being sold with 'vacant possession of the ground floor' but with a Housing Act notice limiting

---

[46] *Secretary of State for the Environment* v *Baylis and Bennett* [2000] 80 P&CR 324.

occupation of the whole house to one household.[47] While these cases would seem clear, case law provides an inconsistent picture as to whether vacant possession can be, and is, given at the relevant time with respect to orders to requisition a property or the service of notices of compulsory purchase.[48]

Requisitioning notices

A small collection of cases concern the government requisitioning of properties under certain provisions of the Defence (General) Regulations 1939. Whilst historic, the decisions are, by analogy, applicable to current cases involving the requisitioning of property under the applicable statute.

The common set of circumstances in cases relating to the requisitioning of properties is that the parties will have entered into written agreements for sale and purchase of a property which then becomes subject to a requisitioning notice. Some case law is clear that a requisitioning notice will not create an encumbrance on the land so as to prevent a seller from giving vacant possession. In *Re Winslow Hall Estate Company* v *United Glass Bottle Manufacturers Ltd*,[49] a contract for the sale of land had been entered into between the parties. Following this, but before completion, notice was given on behalf of the Government to the purchasers that it was intended to requisition the land under the Defence (General) Regulations 1939. The Court held that at the relevant date the vendors were able to give vacant possession. The 'giving' of the requisition notice was held not to create an encumbrance on the land so as to prevent the vendors from performing their contract.

Other authorities suggest that from the moment when the requisition notice was served on the seller, the seller was not in a position to give vacant possession; that is, the notice prevented the giving of vacant possession. In the Court of Appeal decision in *James Macara Ltd* v *Barclay*,[50] the defendant agreed to sell certain property to the plaintiffs. Vacant possession was to be given on completion. Following exchange, but before completion, a government department, as the competent authority under the same Defence (General) Regulations 1939, served the defendant with a notice requisitioning the property. The defendant's solicitors sent a copy of the requisition notice to the plaintiffs; and the plaintiffs subsequently gave notice to the defendant that they rescinded the contract on the ground of the defendant's inability to grant vacant possession. The defendant disputed this and

---

[47] *Topfell* [1979] 1 WLR 446.

[48] A sale contract may provide that it is for the purchaser to comply with outstanding public requirements or that the sale of the property will be subject to such matters as requisitions or compulsory purchases, in which case the risk has already been expressly agreed to have passed to the purchaser or retained by the vendor (as appropriate).

[49] *Re Winslow Hall Estate Company* v *United Glass Bottle Manufacturers Ltd* [1941] Ch 503.

[50] *James Macara Ltd* v *Barclay* [1945] KB 148.

contended that the requisition notice did not, upon its true construction, amount to an exercise of the power to enter into possession under the regulations, and, in fact, no actual entry had been made. It was held that since actual entry was not necessary to exercise the power given by the Regulations, the serving of the requisition notice on the defendant was sufficient to show a present intention to enter into possession of the property. The vendor, therefore, was not, at the date of completion, able to give vacant possession of the property and the first instance decision was affirmed. It was noted in *James Macara* that the notice served was in the same form as that in *Re Winslow Hall*, and so the apparently differing decisions on the same provisions are difficult to reconcile.

Compulsory purchase orders

Cases relating to compulsory purchase orders also provide a confusing picture. Where a compulsory purchase order is made over the property between exchange and completion (or notices are served pursuant to such an order), the question arises as to whether the purchaser may claim that the contract has been frustrated (because vacant possession cannot be given) and that, as a result, the purchaser is not obliged to complete.

In *Hillingdon Estates Co* v *Stonefield Estates Ltd*,[51] the parties agreed to the sale and purchase of certain land. The contract provided that the purchasers were entitled to take up the property in January 1939, and that they should in the conveyance enter into restrictive covenants to the effect that the land should not at any time be used for any purpose other than as a building estate to be developed in a particular manner. The completion of the transaction was delayed, *inter alia*, by the outbreak of war, and in October 1948, when the contract was still uncompleted, the local County Council made a compulsory purchase order affecting the whole of the property. In July 1949, notices to treat under the order were served on the vendors and on the purchasers. The purchasers claimed that, on or before the date of the service of the notices to treat, they were discharged from their contract to purchase the property. It was decided by the Court that the purchasers were treated as owners in equity as soon as a binding contract was made. The service of a notice to treat did not affect the vendors, whose sole interest was to receive the purchase money; it followed that the risk of compulsory purchase properly fell on the purchasers, who were not entitled to rescind the contract. As such, the vendors were in a position to give vacant possession on completion. The same decision was reached in *E Johnson & Co (Barbados) Ltd* v *NSR Ltd* (a decision of the Privy Council),[52] where the publication of a statutory notice warning that land under a contract of sale was likely to be

---

[51] *Hillingdon Estates Co* v *Stonefield Estates Ltd* [1952] Ch 627.
[52] *E Johnson & Co (Barbados) Ltd* v *NSR Ltd* [1997] AC 400.

required for Crown purposes, also did not amount to a frustrating event, meaning that vacant possession could be given by the sellers on completion.

In *Korogluyan* v *Matheou*,[53] the similar question to be decided was whether notices served pursuant to the provisions of the Compulsory Purchase Act 1965, stating that the acquiring authority would be entering upon the land, meant that it was no longer possible for the seller to give vacant possession, and that in consequence the defendant ought to be discharged from his obligation under the agreement. Here, however, it was said (obiter) that, given the nature of the notices served, the plaintiff *was not* in a position to sell with vacant possession in the sense in which the judge felt the words ought sensibly to be construed in the context of the whole transaction. Whilst the seller escaped liability for the payment of damages to the buyer given the contractual conditions of the contract,[54] this case clearly suggested that the compulsory purchase notice *prevented* the seller from being able to deliver vacant possession (contrary to other case law).

Whilst the specific nature of the relevant provisions of the statute in question, and the form of notice served, will have a bearing on the matter, as will the express terms of the legal documentation in question, case law can be seen to have provided an inconsistent position on these issues. Chapter 6 explains the correct position to be taken with respect to these legal obstacles with reference to the model of vacant possession proposed by chapter 5.

Lesser interests

Whilst compulsory purchase orders and requisitioning notices are clearly examples of potential legal obstacles, it is not always clear exactly what is capable of amounting to a legal obstacle to the procurement of vacant possession. The above cases are set in the context of fully fledged rights (or competing claims) to possession, but it is possible to acquire or be granted less extensive rights over land, such as a *profit à prendre* or even the exercise of certain easements or rights of way over a property. Some have questioned if such rights, while amounting to less than possession but still encumbering the estate being transferred in some way, would also be legal obstacles to the receipt of vacant possession.

Imagine that a seller contracts to convey land to a purchaser. The contract provides expressly that vacant possession is to be given on completion. Between exchange and completion, a third party applies to register (and succeeds in

---

[53] *Korogluyan* [1975] 239 EG 649.

[54] See ibid at 649 where it said by Whitford J that '...were it not for the fact that I think the defendant's case fails on special condition 9 and general condition 6, I would have come to the conclusion that at the relevant time the plaintiff was not in a position to sell with vacant possession'. The express conditions in the contract would therefore seem to have had a bearing on the decision.

registering) adverse rights against the property that will prevent development of the land by the purchaser in the manner desired. While the purchaser may have contractual remedies against the seller with respect to disclosure of third party rights, is the seller able to transfer the land to the purchaser on completion in compliance with the seller's obligation to give vacant possession? The third party's rights are clearly an interest over the land rather than a competing claim to possession, but it prevents delivery of the property free from a claim of right over the land (for instance, the right to pass and re-pass) that is adverse to the purchaser. Further, the purchaser may claim that the third party's right to pass and re-pass constitutes (albeit infrequent) third party occupation of the land in some way. The purchaser could clearly argue that the adverse rights were a legal impediment that prevented it from obtaining the quality of possession for which it had contracted.

Chapter 7 seeks to explain the role of lesser interests in cases concerning vacant possession and their difference to the other legal obstacles referred to.

A breach of the obligation?

Given that there is uncertainty as to when an obligation to give vacant possession has arisen, and what any obligation may specifically refer to, it is perhaps not surprising that there are further complications in seeking to ascertain whether a breach of the obligation has taken place. As noted above, historically, decisions as to whether an obligation to give vacant possession had been breached generally proceeded on an ad hoc basis with respect to the particular case in issue. Whilst the specific facts of any particular case will obviously have a bearing on whether a breach of the obligation has arisen (that is, there will also be a case sensitive element to the determination), historically no general principles have been established to ensure consistency and continuity with respect to differing decisions on (ostensibly) similar facts, as high-lighted above. It was not until 1946, in a case concerning rubbish that had been left at the premises, that the Court first laid down what could be seen as a test to determine whether vacant possession had been given.

TESTS FOR BREACH

In *Cumberland*,[55] the plaintiffs contracted to buy a disused freehold warehouse from the defendants. By a special condition, the property was sold 'with vacant possession on completion'. The cellars extending under the whole warehouse were made unusable by rubbish, including many sacks of cement that had hardened. The

---

[55] *Cumberland* [1946] KB 264.

defendant refused after completion to remove the rubbish and the plaintiffs brought proceedings for damages for breach of the condition for delivery of the property with vacant possession on completion. It was held that the defendant had failed to give vacant possession of the property. It was stated that a vendor who leaves his own chattels on property sold by him to an extent depriving the purchaser of the physical enjoyment of part of the property, failed to give vacant possession. Such acts were consistent with the vendor seeking to continue to use the premises for his own purposes,[56] rather than passing possession to the purchaser in accordance with the terms of the contract. It was further noted that it was no answer for the vendor to claim to have abandoned his ownership of the chattels on completion to prevent a breach of the obligation. The Court held that 'the existence of a physical imped-iment, which substantially prevented or interfered with the enjoyment of the right of possession of a substantial part of the property, to which the purchaser did not expressly or impliedly consent to submit ...'[57] would cause a breach of the vacant possession obligation. This was qualified as being subject to a *de minimis* rule, even though no specific details as to the nature and form of that rule were elaborated upon. The *Cumberland test* (as it is known) which remains the authority, was further elaborated upon in recent years in the context of the procurement of vacant possession when exercising a contractual break option in a lease.

In *John Laing Construction Ltd* v *Amber Pass Ltd*,[58] the claimant was the tenant of commercial premises under a lease granted by the defendant's predecessor-in-title. A clause in the lease provided that the lease might be determined by *inter alia* the 'yielding-up of the entirety of the demised premises'. The claimant sought a declaration that, pursuant to a notice given under the break clause, it had validly terminated the lease. That claim was contested by the defendant, which sought to counter-claim for declarations that the purported break notice was ineffective and that the lease was therefore continuing. The defendant contended that the claimant had not 'yielded-up' the property, relying, *inter alia*, on the continued presence of security guards at the premises and the claimant's failure to hand back the keys to the premises. The defendant contended that these were inconsistent with providing vacant possession at the end of the term.

---

[56] The decision in *Lysaght* v *Edwards* [1876] 2 Ch D 499 discussed the use of the premises as a dump for one's own purposes or for leaving there that which substantially prevented or interfered with the enjoyment of possession of a substantial part of the property. See also *Norwich Union Life Insurance Society* v *Preston* [1957] 2 All ER 428, where a mortgagor which had left furniture in the premises after a court order requiring him to give up possession had not complied with the law and was using the premises for his own purposes as a place for the storage of his goods.

[57] *Cumberland* [1946] KB 264, per Lord Greene at 269.

[58] *John Laing Construction Ltd* v *Amber Pass Ltd* [2004] All ER (D) 115 (Apr).

The Court found for the claimant and held that it had plainly and obviously manifested a desire to terminate the lease and was accordingly entitled to the declaration sought. The continued presence of security guards at the premises and the tenant's failure to hand back the keys had not prevented vacant possession being given. The Court held that the task of the Court was:

> to look objectively at what had occurred and determine whether a clear inten-
> tion had been manifested by the person whose acts were said to have brought
> about a termination to effect such termination, and whether the landlord
> could, if it wanted to, occupy the premises without difficulty or objection.[59]

The decisions in *Cumberland* and *John Laing* can therefore be seen to have developed a two-limb test to determine a breach of the obligation to give vacant possession. The first limb focuses on the intention of the party required to give vacant possession (as manifest by its conduct in purportedly vacating the premises), with the second looking at whether the party with the right to vacant possession can, at the point of completion, occupy the property (or a substantial part of the property) without difficulty or objection.

PROBLEMS WITH THE TESTS

Whilst of some assistance to lawyers and other practitioners, the tests that case law has developed would appear far from conclusive, especially with respect to the second limb.

For example, it is unclear what extent of 'difficulty' under the wording provided by the *John Laing* decision is required, and whether this must be general inconvenience or significant distress. *Cumberland* suggests that a tenant has to remove all chattels and also rubbish which 'substantially prevents or interferes with enjoyment of a substantial part of the property', but there is no definition of what constitutes 'substantial' or whether this test is purely objective or whether there is a subjective element to it. Indeed, unlike the first limb of test which is clearly directed at the intention of the party required to give vacant possession as manifest by its conduct, there is some debate as to whether the second limb of the test should be judged against any purchaser or landlord seeking to (re)take possession, or against a purchaser or landlord with particular qualities of the actual purchaser or landlord in question. It can therefore be questioned as to whether, for example, the court should consider more generally whether rubbish left at the property, on the break of a lease or completion of a sale, prevents the average purchaser or landlord (objectively speaking) from (re)occupying without difficulty or objection or (objectively

---

[59] *John Laing Construction Ltd* v *Amber Pass Ltd* [2004] 2 EGLR 128, per Robert Hildyard QC at 131.

speaking) the actual purchaser or landlord in question given its specific circumstances. As such, it is arguable that specific circumstances (and intended use for the premises) of the purchaser (or landlord) is relevant to the so-called 'objective' second limb of the test, just as the intention of the party seeking to give vacant possession (as demonstrated by its conduct) is material to the first limb of the test.

It is also unclear as to whether there has to be an actual interference, or whether the likelihood or potential for the leftover items to cause a substantial interference will be sufficient. Further, it has not been clarified as to what counts as a valid objection and again, whether specific factors relating to the purchaser or landlord in question can be taken into account when determining whether the purchaser or landlord can (objectively speaking) (re)occupy without such an objection. This raises the likelihood that a materially similar objection could be deemed valid in one context, but not in another, given the specific circumstances of the parties in question, and the parties' intended use of the premises from completion.

It must also be borne in mind that this confusion is in the context of physical and tangible items which are claimed to have caused a breach of the vacant possession obligation. It would be even more problematic to seek to apply the tests laid down to a situation where, for example, it was claimed that the difficulty or objection related to the state of the premises (which may, for example, have been destroyed by fire immediately before completion). As noted earlier, there remains no actual authority on the position where the seller's inability to give vacant possession is due to the physical state of the property,[60] and how the tests could possibly be applied in such a context. The specific nature and form of any potential legal impediment to vacant possession could also render the tests relatively ineffective in certain contexts. It is therefore questionable whether the current tests are of any assistance at all in determining whether vacant possession has been given at the relevant time in a wide variety of everyday circumstances, as discussed in further detail in chapter 7.

The consequence of this current inadequate situation is that the parties in any given case will be unsure as to their legal rights and obligations and accordingly unable to determine, with certainty, how they should seek to resolve the situation. This potentially puts a greater burden on the party with the weaker financial strength and resource which may be unable, or unwilling, to litigate on a given dispute that may arise.

Chapter 5 discusses both limbs of the tests in greater detail in order to seek to explain how the respective limbs are likely to be interpreted as operating in practice.

---

[60] According to Harpum, 'Vacant Possession − Chamaeleon or Chimaera?' (1988) *Conveyancer and Property Lawyer* 324, 400.

Remedies for Breach

On the basis that an obligation to give vacant possession has arisen and is breached by the party required to give vacant possession, it must be considered where this leaves the party which had contracted for something more than is actually obtained at the relevant time.

It has been argued that if vacant possession is a term of a contract (either expressly or impliedly) and between exchange and completion some supervening event makes it impossible for the seller to give vacant possession to the purchaser on completion, then the contract may be deemed frustrated. As noted above, examples include cases relating to compulsory purchase orders[61] and the requisitioning of a property for specific purposes.[62] Frustration is very rarely claimed however, and case law appears to have shown unwillingness, by the courts, to find that a contract has been frustrated.[63]

OPTIONS IN THE EVENT OF BREACH

If, on the day of completion (but before completion is effected), a purchaser was to inspect the premises and see that the premises were not vacant, it could do the following:

1. Apply to the court for an order for specific performance,[64] and claim damages.
2. Serve a notice to complete on the seller and after expiry of that notice (which will be determined by contractual provisions) rescind the contract, recover any deposit paid and claim damages.
3. Choose to complete without prejudice to a right to claim damages.[65]

Whilst it is better to inspect a property immediately before completion, in the majority of cases a given property is not inspected prior to completion.[66] The first a purchaser knows about the problem with vacant possession is after completion when it arrives at the premises to find that all is not as it had expected. At this point,

---

[61] *Korogluyan* [1975] 239 EG 649.

[62] *Cook* [1942] 2 All ER 85; *James Macara* [1945] KB 148.

[63] *Hillingdon Estates Co v Stonefield Estates Ltd* [1952] applied by Privy Council in *E Johnson & Co (Barbados) Ltd v NSR Lt*d [1997] AC 400.

[64] According to *Wroth v Tyler* [1964] Ch 30, a seller will not normally be obliged by an order for specific performance to undertake 'hazardous' litigation to obtain possession, but would still remain liable in damages.

[65] The availability and amount of damages will depend on the circumstances, the nature of the losses and the express terms of the contract.

[66] See Lexis Nexis Butterworths Document [547] 10 Occupiers (accessible via subscriber service).

the contract has been completed (the seller has the sale monies in cleared funds) and the purchaser is left having to claim damages for a property that it cannot immediately occupy as it wished to. This is hugely unsatisfactory. Further, it leaves the purchaser with the burden of having to advance a claim to recover the loss sustained as a consequence of the breach of the vacant possession obligation and this may prove difficult, or impossible if, for example, the seller has weak covenant strength. If the obstacle to vacant possession is a person or entity with a statutory right to remain in occupation, the purchaser may have difficulty in removing that person or entity from the property or be required to take the property subject to its interest based on the doctrine of 'constructive notice', which deems a party with having knowledge which it did not in fact actually have.[67]

POST COMPLETION ACTION

The breach of an obligation to give vacant possession gives a purchaser the right to choose to action the breach even *after* completion. This is because the obligation to give vacant possession has been said not to merge in the conveyance or transfer, but to remain actionable after completion (even in the absence of an express non-merger clause). It is commonplace for purchasers to seek to action a breach of the obligation to give vacant possession after completion by suing the purchaser for damages which, as discussed in chapter 8, will be determined by the nature of the impediment which prevented the procurement of vacant possession on completion. Further, there is authority for the proposition that a purchaser will be able to *terminate* the contract even after completion has taken place (effectively unravelling the contract and having the purchase monies returned to it), subject to the purchaser having not affirmed the contract.[68] However, such a course of action is significantly less commonly undertaken, most likely because, once transferred, it may be very difficult to recover the purchase monies.

PRACTICAL ISSUES

It is apparent that, at present, the current law and practicalities of completion put the seller in a much stronger position as far as a breach of a vacant possession obligation

---

[67] For a discussion of the doctrine (with respect to overriding interests and other adverse interest to which a sale may be subject), see Howell, 'Notice: A Broad View and a Narrow View' (1996) Conv 34; Partington, D 'Implied Covenants for Title in Registered Freehold Land' (1988) Conv 18 and Sheridan, D 'Notice and Registration' (1950) NILQ 33. In respect of issues concerning title, registration has sought to obviate constructive notice considerations in recent years by protecting purchasers from the undiscoverable interests in their otherwise clean title.

[68] See *Hissett* v *Reading Roofing Co Ltd* [1969] 1 WLR 1757 and *Gunatunga* v *Dealwis* [1996] 72 P&CR 161.

is concerned. A purchaser, having paid the monies over only to find that it cannot immediately occupy without difficulty or objection, will often be left in the difficult position of advancing a claim for damages against the seller. The purchaser can sometimes suffer even greater detriment if it has already contracted to demise the premises to a tenant on the basis that a transfer to it takes place (or has entered into a contract to immediately sell on the property). This could result in the purchaser itself potentially being subject to a breach of contract claim (with respect to an anticipated tenant or sub-buyer), giving rise to consequential losses.

Chapter 8 discusses the issue of remedies in more detail and proposes some enhanced contractual provisions relating to damages which may more evenly distribute power between parties, and more fairly apportion their respective obligations and responsibilities.

Summary

It is clear from the above that problems are manifest at every stage with respect to the obligation to give vacant possession. Even if one can determine that an obligation has arisen (expressly or impliedly), what this will actually refer to is likely to be unclear, and how a breach can be established will potentially be difficult to ascertain. Even then, remedies flowing from any breach established may be inadequate or unsatisfactory. Therefore at every stage, the issues of risk and responsibility for the parties in question can be seen to be a major issue and pervasive to all aspects of the obligation to give vacant possession.

In order to seek to suggest how the obligation could be better interpreted, it is first necessary to fully evaluate how it has been understood (or misunderstood) over time. As such, chapter 3 develops the notion of vacant possession as an express and implied term of the contract, by examining how vacant possession terms have interacted with other contractual terms, and what this reveals about the overriding status of the obligation to give vacant possession in the context of standard sale and purchase contracts. This helps to explain the precedence of a term for 'vacant possession' when appearing as a special condition of the contract, and the dangers of not providing for vacant possession in such a way.

Chapter 3

Vacant Possession as a Contractual Condition

In chapter 2, the current problems with the obligation to give vacant possession were highlighted. It was shown that a number of uncertainties surround the very essence of the obligation in terms of when it may have arisen, what it refers to, how it may have been breached and what remedies will be available for a breach.

This chapter considers the role of vacant possession as a term of a contract by explaining its interaction with other terms (or conditions). The chapter focuses first on decisions relating to conflicts between express clauses providing for vacant possession, when appearing as a *special* condition, and other contractual conditions. It then discusses how that position differs where other terms conflict with an express *general* (as opposed to special) condition for vacant possession, or when other terms conflict with only an *implied* obligation for vacant possession. Whilst the precedence of an express special condition for vacant possession is shown to have been established by case law, a lack of authority continues to leave the position unclear in cases where the vacant possession condition is only a general condition, or is merely implied into the contract.

Vacant Possession: Law and Practice. ISBN: 978-0-08-096680-9

Express vacant possession clauses

The precise terms of the sale and purchase contract are key to understanding the issue of vacant possession:

> *What the purchaser is entitled to get in the way of possession on completion depends, of course, on what the contract says.*[1]

As noted in chapter 1, the obligation to give vacant possession normally appears expressly as a term in a legal agreement, conveyance, contract or transfer. In practice (and since 1902), contracts for the sale of land have incorporated standard conditions of sale by reference (as an appendix to the contract) and these have included conditions for vacant possession. These standard conditions of sale (which are incorporated as terms of the contract)[2] ultimately determine the parties' rights and obligations under the contract, and remedies in the event of a breach by either party. The various editions and versions of the conditions of sale each set out the 'general' and 'special' conditions of the sale and purchase agreement. The general conditions, which evolved throughout the twentieth century, deal with various issues relevant to the sale and purchase of property, for example, insurance, deposits, requisitions and matters relevant to completion. Special conditions highlight specific aspects of the transaction especially of importance to the parties and provide an opportunity to address any unique factors relevant to the transaction that the general conditions do not adequately cater for. Unlike the general conditions, which are a standard printed set of conditions attached to the main contract, special conditions are manually written or typed on a separate page. As discussed in more detail in chapter 4, it is established in the conditions of sale that special conditions should take precedence over any inconsistent general conditions.

POSSIBILITY FOR CONFLICT WITH OTHER CONDITIONS

A special condition (or express statement in the particulars of sale)[3] that requires vacant possession to be given on completion, is obviously something that the parties will have specifically considered before recording expressly in their agreement. Such a condition is likely to provide that:

> *The Property is sold with vacant possession on completion.*

---

[1] Farrand, JT *Conveyancing Contracts: Conditions of Sale and Title* (London, Oyez Publications, 1964) p 259, under the section entitled 'Vacant Possession'.

[2] Cheshire, GC and Burn, EH *Modern Law of Real Property* (London, Butterworths, 4th ed, 1976) p 74. See chapter 4 for further details.

[3] Special conditions are deemed to include the terms of the particulars of sale, see for example the *National Conditions of Sale* (London, The Solicitor's Law Stationery Society Ltd, 20th ed, 1981).

Whilst it was common for contracts to include such an express special condition for vacant possession, it is also commonplace for such contracts to include other clauses within the agreement that may have reference to, or an effect on, the issue of vacant possession. For example, a contract for the sale and purchase of land is likely to include a clause dealing with a seller's liability for errors and omissions in the contract. If the contract wrongly provides that vacant possession is to be given on completion, then a seller may seek to rely on the 'errors and omissions' clause to claim that the buyer cannot insist on vacant possession, as provided for by the contract, because the clause providing for vacant possession was an 'error'. When such a clause is potentially at odds with the term for vacant possession, a conflict can arise in respect of how such a condition should be interpreted as modifying (or altering) the vacant possession clause.

When a conflict arises between a term relating to vacant possession and another contractual term, it is ultimately the role of the courts to determine which condition should prevail and which should be subordinate. Understanding how the courts have dealt with conflicts of this kind over time therefore assists in understanding the nature and form of the vacant possession obligation, and the established precedence of the term in cases where it is incorporated into the contract as a special condition.

INTERPRETATION OF CONFLICTING PROVISIONS

Historically, purely 'mechanical' type rules were employed to resolve such contradictory provisions of contractual documents. In *Forbes* v *Git*,[4] Lord Wrenbury said that:

> *...if in a deed an earlier clause is followed by a later clause which destroys altogether the obligation created by the earlier clause, the later clause is to be rejected as repugnant and the earlier clause prevails.*[5]

In recent years, a series of decisions[6] has led to 'a fundamental change'[7] in the approach taken by the courts to the interpretation of documents of all kinds. The principle lying behind the modern approach to the interpretation of documents is that

---

[4] *Forbes* v *Git* [1922] 1 AC 256, per Lord Wrenbury (PC) at 259.

[5] ibid at 259.

[6] In particular, see *Prenn* v *Simmonds* [1971] 1 WLR 1381; *Reardon Smith Line* v *Hansen-Tangen* [1976] 1 WLR 989; *Charter Reinsurance* v *Fagan* [1997] AC 313; *Mannai Investment Co Ltd* v *Eagle Star Life Assurance Co Ltd* [1997] AC 749; *Investors Compensation Scheme Ltd* v *West Bromwich BS* [1998] 1 WLR 896; *Bank of Credit and Commerce International SA* v *Ali* [2002] 1 AC 251 and *Sirius International Insurance Co* v *FAI General Insurance* [2004] 1 WLR 3252.

[7] *ICS Ltd* v *West Bromwich BS* [1998] 1 WLR 896, per Lord Hoffmann at 912.

the meaning that should be attached to particular words is heavily dependent upon the context in which those words have been used. The Court is normally seen to try to give effect to every clause in the contract,[8] seeking to interpret each (so far as is possible) in order:

> ...*to bring them into harmony with the other provisions of the [contract], if that interpretation does no violence to the meaning of which they are naturally susceptible.*[9]

This will often lead to the court concluding that one clause qualifies another in some way. A clause will be rejected as being contrary to the tenor of the agreement if there really is no alternative.[10] Normally, in the event that there are two competing or conflicting clauses in a contract, the Court will have to make a determination as to which is the leading provision and which must be viewed as subordinate and only able to be given meaning to the extent that it does not contradict the dominant clause.[11]

Express (special conditions) for vacant possession

Over time, a number of judges have had to rule on the interaction between an express obligation to give vacant possession (when appearing as a special condition) and other contractual terms. The outcomes of some of these decisions have led to criticism[12] and subsequently been overruled (or not followed).[13] It is appropriate to review two types of contractual condition that have been found to interact with an express special condition for vacant possession to elucidate how it has not always been clear that the vacant possession obligation should take precedence (which is the correct position, where the vacant possession clause appears as a special condition). As discussed, both the decisions analysed in the sections which follow were therefore incorrect, as explained in the section entitled 'Precedence of the obligation'.

---

[8] *Chitty on Contracts* (London, Sweet & Maxwell, 25th ed, 1983) p 429, para 784.

[9] *Chamber Colliery Co Ltd* v *Twyerould* [1915] 1 Ch 268, per Lord Watson HL at 272.

[10] *Forbes* [1922] 1 AC 256, per Lord Wrenbury at 259.

[11] *Halsbury's Laws of England* (London, Butterworths, 4th ed, 1983) Vol 44, para 872; *Institute of Patent Agents* v *Lockwood* [1894] AC 347, per Lord Herschell LC at 360. See also Harpum, C 'Vacant Possession - Chamaeleon or Chimaera?' (1988) *Conveyancer and Property Lawyer* 324, 400.

[12] See Harpum, 'Vacant Possession - Chamaeleon or Chimaera?' (1998) *Conveyancer and Property Lawyer* 324, 400 and Barnsley, DG 'Completion of a Contract for the Sale and Purchase of Land: Part 3' (1991) *Conv* 185 at 188.

[13] See Templeman J in *Topfell Ltd* v *Galley Properties Ltd* [1979] 1 WLR 446.

'No Annulment, No Compensation' Clauses

As referred to above, the first specific context in which a conflict has been seen to arise is with respect to so-called 'no annulment, no compensation' clauses. Clauses like these are likely to state that:

> *no error, misstatement or omission in the particulars, sale plan or conditions shall annul the sale, nor shall any compensation be allowed either by the vendor or the purchaser in respect thereof.*[14]

If a seller expressly contracts to sell land with vacant possession but then finds that it cannot, the effect of a general condition of this kind on the express obligation given by the seller as to vacant possession must be considered. In theory, if the impediment amounted to an error, misstatement or omission from the particulars of sale, a general condition of this kind may have an effect on the express special condition as to vacant possession. The seller may argue that it is not in breach of its obligation to give vacant possession by virtue of the 'no annulment, no compensation' clause, which has the effect of preventing the sale from being declared void as a result of an error or misstatement. The effect of this would be that the purchaser is required to take the property subject to the impediment, even though the purchaser believed that it was contractually entitled to vacant possession on completion (something, which the purchaser is now not obtaining). In other words, the clause may enable the seller *not* to give vacant possession (as has been expressly contracted for) but nevertheless to escape liability for breach by relying on such a 'no annulment, no compensation' clause in the contract.

This question arose in *Curtis* v *French*[15] where the defendant contracted to sell a cottage to the plaintiff. This decision highlights how the courts have previously taken an incorrect approach to this issue.

The contract incorporated the *National Conditions of Sale* (10th edition) and included, as condition 10, a 'no annulment, no compensation' clause as a *general* condition which provided that:

> *No error mis-statement or omission in the particulars . . . shall annul the sale nor shall any compensation be allowed either by the vendor or purchaser in respect thereof.*[16]

---

[14] For example, see the *National Conditions of Sale* (London, The Solicitor's Law Stationery Society Ltd, 10th ed, 1927) condition 10. Similar provisions are currently included in the general conditions of the *Standard Conditions of Sale* (London, The Law Society, 4th ed, 2003) condition 7.1 and *Standard Commercial Property Conditions* (London, The Law Society, 2nd ed, 2003) condition 9.1.

[15] *Curtis* v *French* [1929] 1 Ch 253, per Eve J.

[16] ibid at 260.

The particulars of sale included an express *special* condition as to vacant possession and stated that the property was let to:

> *a local farmer, for one of his employees. The tenant formally and legally terminated the tenancy, but has not yet handed over vacant possession, the vendor has not yet pressed for possession, allowing the occupier to remain on sufferance, but the premises will be sold with vacant possession.*[17]

The statement in the particulars was materially incorrect. The local farmer had not terminated the subtenancy and the subtenant was actually a statutory protected tenant in any event under the Increase of Rent and Mortgage Interest (Restrictions) Act 1920. The seller had unsuccessfully sought possession for some time but had not commenced any proceedings to evict the occupier. Accordingly, the Court was charged with determining whether the seller could be allowed to rely on the 'no annulment, no compensation' clause when the purchaser apparently 'affirmed' the contract and sought damages for the seller's failure to give vacant possession.

Eve J decided (wrongly) that reliance *could* be placed on the 'no annulment, no compensation' clause by the seller, even though the seller was fully aware that the presence of the tenant would prevent him from giving vacant possession and that he failed to adequately disclose this to the purchaser. In accordance with established principles and case law, reliance on the clause should not have been permitted per se because of the non-disclosure.[18] Specifically with reference to the issue of the procurement of vacant possession, counsel for the purchaser argued that:

> *The vendor cannot use [the 'no annulment, no compensation' clause] for the purpose of converting the express promise that the premises will be sold with vacant possession on completion into the exact opposite.*[19]

Whilst both these points were advanced in argument, neither was even referred to in the judgment. Eve J did, however, show an appreciation of the hierarchy of special and general conditions and the extent to which one type of condition should take precedence over another. Here, however, the judge considered that the interpretation of the *general* 'no annulment, no compensation' condition in this context was such that it should override the *special* condition for vacant possession, which should thus be subject to the general condition.

---

[17] ibid at 260.

[18] See *Flight* v *Booth* [1834] 1 Bing (NC) 370 in respect of misdescription and *Re Puckett and Smith's Contract* [1902] 2 Ch 258 CA concerning non-disclosure. Also *Nottingham Patent Brick and Tile Co* v *Butler* [1885] 15 QBD 261 at 271, per Wills J; subsequently applied by Millett J in *Rignall Developments Ltd* v *Halil* [1988] Ch 190 at 197–198.

[19] *Curtis* [1929] 1 Ch 253 at 256, per Ronald Roxburgh for the purchaser.

Eve J appeared to justify such a conclusion by reading into the general 'no annulment, no compensation' clause the reference to 'each lot is sold subject to all tenancies' (from special condition 8 of the contract). This was even despite the misrepresentation by the seller in the particulars of sale as to the tenant remaining in the property on sufferance and the seller not having pressed for possession.

The effect of the decision of Eve J was to render the seller's express undertaking to give vacant possession (a special condition of the contract) effectively worthless and redundant, as the interpretation applied by the Court took away from the purchaser the right to any remedy for a breach of the condition.[20] The decision effectively subordinated the special condition for vacant possession in favour of the relevant general condition of the contract. In doing so, the decision deprived the buyer of the right to receive the property in the manner in which he believed he had contracted with the seller,[21] thus undermining the nature and effect of the special condition for vacant possession in the contract. Given Eve J's interpretation, there was, in essence, little point in the contract including a special condition for vacant possession. As shown in the section entitled 'Precedence of the obligation' below, this decision could not therefore be held to be correct or fair, in similar terms to the decision in *Korogluyan* v *Matheou*[22] which is discussed below.

'SUBJECT TO LOCAL AUTHORITIES' REQUIREMENTS' CLAUSES

A second type of conflicting contractual condition which elucidates how the courts have had to determine whether the term for vacant possession or the conflicting term should take precedence, and how the terms could be held to modify or interpret each other, is a 'subject to local authority requirements' clause. Again, the courts can be seen to have previously made incorrect decisions as to whether the vacant possession clause should take precedence.

The sort of clause in issue here commonly provides that a purchaser is required to accept the property on completion, subject to any notices served in respect of the property. Examples of such notices would include compulsory purchase orders, Housing Act stipulations as to occupation or even tree

---

[20] The 'no annulment, no compensation' clause could obviously still be applicable in the overall context of the contract in respect of matters *not* affecting the express obligation to give vacant possession.

[21] See the later decision of *Sharneyford Supplies Ltd* v *Edge* [1987] Ch 305. Here, quite correctly, a seller who failed to give vacant possession because of the presence of a tenant (whom he had taken no steps to evict), was found to have breached the obligation to give vacant possession.

[22] *Korogluyan* v *Matheou* [1975] 30 P&CR 309.

preservation orders. Such a clause would typically provide that the purchaser is to take the land:

> subject to all notices, orders or requirements given, made or required by the local or other authorities.[23]

Whitford J considered a 'subject to ...' clause in the decision in *Korogluyan*.[24] This case has striking parallels, in principle, to the decision in *Curtis* given that the Court was also required to rule on the interaction between an express clause (appearing as a special condition) providing for vacant possession and other conflicting conditions of sale and, as discussed below, also gave precedence to the other conditions. Both decisions also made similar (and yet fundamental) errors based on their apparent disregard of the non-disclosure of burdens by the respective sellers.

In *Korogluyan*, the property was sold at auction with the particulars of sale stating that vacant possession would be given on completion (that is, there was an express contractual obligation to procure vacant possession, equivalent to a special condition). The contract of sale incorporated a 'subject to ...' general condition (as general condition 6):

> Each purchaser shall be deemed to purchase with full knowledge of the state of repair of the lot or lots purchased by him and of the tenancies thereof (if any) and shall be responsible for all repairs including sanitary requirements and all requirements of the lessor local or other authorities. The properties are sold subject to all notices, orders or requirements whether referred to in the particulars or not, given, made or required by the local or other authorities. Each property shall as from the date of the contract be at the sole risk of the purchaser thereof.[25]

The contract also contained a *special condition* (number 9) that the purchaser was buying with full knowledge of burdens and requirements for the property. It provided that the purchasers:

> having had the opportunity of making all appropriate inquiries of the local authorities shall be deemed to purchase with full knowledge of all entries on the registers kept by them and of all their requirements or proposals

---

[23] Local authority requirement clauses are common to the various editions and revisions of the conditions of sale and are incorporated into current editions as general conditions. For example, see *Standard Conditions of Sale* (London, The Law Society, 4th ed, 2003) condition 3.1.2(e) and *Standard Commercial Property Conditions* (London, The Law Society, 2nd ed, 2003) condition 3.1.2(e).

[24] *Korogluyan* [1975] 30 P&CR 309.

[25] ibid at 315.

relating to the property and shall raise no objection or requisition whatsoever in respect of or in relation thereto.[26]

Before the sale, the local authority served a notice to enter on the seller pursuant to a compulsory purchase order, which was not revealed at auction. As such, on completion it was not possible for the seller to give vacant possession as had been stipulated in the particulars of sale.

In his judgment, Whitford J held that the seller was able to rely on the general and special conditions referred to above. Although the auction particulars clearly said that vacant possession would be given on completion, according to the judge, general condition 6 and special condition 9 of the agreement 'put the purchaser to inquiry' as to local authority requirements and possible notices affecting the property. He considered that the effect of those conditions was to preclude the possibility of any complaint that, as a result of the service of any notice by the local authority, it became impossible for the plaintiff to give vacant possession:

> *general condition 6 and special condition 9...do draw the attention of the purchaser to the fact that it may be sensible to see what the position* vis-à-vis *any local authority requirements may be in relation to this particular property, and do draw the purchaser's attention to the fact that if notices may have been served, then the purchase is going to be effected subject to such burdens as the notices given may place upon the property in question.*[27]

Whitford J assumed that the conditions sufficiently alerted the purchaser to the risk that there might be a compulsory purchase order, and treated this as sufficient to qualify the seller's express undertaking to give vacant possession. His decision was therefore that the express obligation to give vacant possession was qualified by these conditions.[28]

Given his knowledge of the compulsory purchase order at the time of the auction, but corresponding lack of disclosure, in a similar manner to the seller in *Curtis*, the seller should not have been able to rely on the (general) 'subject to ...' condition as a matter of law. This is because a vendor who knows or ought to have known of such a notice, order or requirement, cannot rely on such a condition if the burden is not disclosed at the time of contract.[29] Further, as outlined in chapter 2, case law has since confirmed that a seller's *express* obligation to procure vacant possession has the effect of making a purchaser's knowledge of *any* impediment (removable or

---

[26] ibid at 315.

[27] ibid at 317.

[28] ibid at 317.

[29] *Nottingham Patent Brick and Tile Co* v *Butler* [1885] 15 QBD 261, per Wills J at 271; subsequently applied by Millett J in *Rignall Developments Ltd* v *Halil* [1988] Ch 190 at 197−198.

irremovable) to vacant possession immaterial.[30] As such, even if the purchaser could be *deemed* to have had knowledge of the risk of such a notice being served (as the judge suggested), that impediment to vacant possession should have been of no consequence in the light of the express condition for vacant possession.[31] Due to the seller expressly contracting to sell the property with vacant possession, special condition 9 was completely irrelevant as far as the obligation to give vacant possession was concerned.

Moreover, the seller should not have been allowed to qualify his express undertaking to give vacant possession by other conditions in any event, otherwise the express vacant possession term was being undermined. The express condition that the seller will provide vacant possession on completion, as an integral part of the contract, ought to have prevailed over such a 'subject to ...' clause, and the other conditions of sale should not have been allowed to negate the express undertaking as to vacant possession. Indeed, as previously noted, the various versions of the conditions of sale themselves confirm that special conditions (of which vacant possession will normally be one) have priority over any inconsistent general provisions. The special conditions provide that the general conditions apply so far as they are not varied by or inconsistent with these special conditions. For example, under the Law Society's *Contract for Sale* (1984 Revision), special condition A provides that the general conditions apply 'so far as they are not varied by or inconsistent with these special conditions'.[32] The National Conditions of Sale provided that the general conditions apply so far as they are 'not inconsistent' with the special conditions.[33] This was an established principle from the conception of the Conditions of Sale,[34] and remains the case in the current editions of the *Standard Conditions of Sale* and *Standard Commercial Property Conditions* (see chapter 4).

The judge made no reference to the decision in *Curtis* even though this actually supported, in principle, the conclusion reached in *Korogluyan*, namely that the seller could rely on other conditions of sale to effectively 'convert' the express promise that the premises would be sold with vacant possession on completion into the exact opposite. As such, in both these decisions, the sellers were (wrongly) able to escape

---

[30] *Sharneyford Supplies Ltd* v *Edge* [1987] Ch 305.

[31] ibid. It should also be noted that the purchaser was not actually objecting to the compulsory purchase order in its own right, but to the seller's failure to provide vacant possession pursuant to his express obligation to do so in the relevant particulars of sale. See also *Phillips* v *Caldcleugh* [1868] LR 4 QB 159 (in the context of disclosures on title).

[32] *The Law Society's Contract for Sale* (1984 Revision) (London, The Law Society, 1984).

[33] See, for example, the *National Conditions of Sale* (London, The Solicitor's Law Stationery Society Ltd, 19th ed, 1976).

[34] See [1953] 97 *Sol Jol* 395.

liability for not giving vacant possession by virtue of the other contractual terms. As discussed below, these decisions were erroneous and flawed.

PRECEDENCE OF THE OBLIGATION

The decisions in *Curtis* and *Korogluyan* undermined the purpose of an express special condition for vacant possession. They manifested a disregard to the precedence of an obligation for vacant possession and for the hierarchy between special conditions and general contractual conditions. These decisions, negating the effect of an express promise as to vacant possession, stood as authorities until 1979 and the decision of Templeman J in *Topfell Ltd v Galley Properties Ltd*.[35] This provided the first coherent statement of what a special condition for vacant possession was seeking to provide in the context of a contract for the sale and purchase of land, and what place it has with reference to other incorporated conditions.

In *Topfell*, the purchaser acquired a property from the defendant with the particulars of sale stating (in bold type) that the property was sold with vacant possession of the ground floor.[36] As such, it was an express special condition of the contract that vacant possession would be given. Prior to the sale, the local authority served a notice under section 19 of the Housing Act 1961, directing that the house was to be occupied by only one household. A pre-existing tenancy of the first floor thus precluded occupation of the ground floor by another household. Vacant possession of the ground floor could not therefore be given on completion contrary to the express contractual provision that vacant possession would be provided.

The sellers knew of this notice but did not disclose it to the purchaser. The purchaser successfully sued for specific performance with an abatement of the price. The contract contained a 'subject to ...' clause (which was a general condition) and identical to that which was in issue in *Korogluyan*:[37]

> *Each purchaser shall be deemed to purchase with full knowledge of the state of repair of the lot or lots purchased by him and of the tenancies thereof if any and shall be responsible for all repairs including sanitary requirements and all requirements of the lessor, local or other authorities. The properties are sold subject to all notices, orders or requirements, whether referred to in the particulars or not, given, made or required by the local or other authority. Each property shall as from the date of the contract be at sole risk of the purchaser thereof.*

---

[35] *Topfell Ltd* v *Galley Properties Ltd* [1979] 1 WLR 446, per Templeman J at 450.
[36] ibid at 450.
[37] *Korogluyan* [1975] 30 P&CR 309.

The sellers relied on this in their defence, along with a 'no annulment, no compensation' clause (which was a special condition of the contract) and practically the same as that found in the decision in *Curtis*:[38]

> *The purchaser having had the opportunity of making all appropriate inquiries of the local and other authorities shall be deemed to purchase with full knowledge of all entries on the registers kept by them and of all their requirements or proposals relating to the property and shall raise no objection or requisition whatsoever in respect of or in relation thereto.*[39]

In a similar manner to *Korogluyan*,[40] it was argued in submissions that if the purchaser had in fact made local searches and inquiries, the direction imposed under the Housing Act 1961 would have been revealed. In other words, it was argued that the purchasers had notice of the risk of such notices and should therefore take the property subject to that legal obstacle to the receipt of vacant possession, in accordance with the aforementioned general and special conditions.

Templeman J was very clear, however, as to how the interaction between the express special condition for vacant possession and these apparently contradictory general and special conditions should be dealt with. He said:

> *...these special and general conditions cannot be allowed to contradict the contractual obligation into which the [sellers] entered ... to give vacant possession.*[41]

As a matter of construction, Templeman J regarded the obligation to give vacant possession as the paramount provision, and the other special and general conditions of sale as subordinate to it. *Topfell* therefore confirmed that an express undertaking to give vacant possession constitutes an overarching guarantee by the seller (i.e. an express vacant possession obligation (appearing as a special condition) is the leading provision). Accordingly, *Topfell* held that any condition of sale which purports to modify that obligation should be construed as being subordinate to the express undertaking. This was seen to apply against general and other special conditions of the contract, with the express special condition for vacant possession taking precedence.[42]

---

[38] *Curtis* [1929] 1 Ch 253, per Eve J.

[39] *Topfell* [1979] 1 WLR 446.

[40] *Korogluyan* [1975] 30 P&CR 309.

[41] *Topfell* [1979] 1 WLR 446, per Templeman J at 450.

[42] Interestingly, no reference was made to the earlier 'authorities' on this point, thus leaving it unclear as to whether such decisions were overlooked or disregarded.

Harpum, discussing these decisions, concluded that the decision in *Topfell* was to be preferred, and stated that:

> *An express undertaking to give vacant possession constitutes an* overriding guarantee *by the vendor.* Any *condition of sale which would otherwise have limited the obligation should in general be construed as being qualified by the express undertaking and not vice versa.*[43]

Whilst not explicitly mentioning why the decision in *Topfell* was to be regarded as the most correct, and simply stating that 'it is respectfully suggested that this conclusion is the right one',[44] a number of justifications can be provided. These further support the claim that *Topfell* is the better decision (even though the decision in *Topfell* never formally overruled the earlier decisions in *Curtis* and *Korogluyan*).

First, as alluded to by Harpum, the decision in *Topfell* supports the hierarchy that has long been established between special and general conditions of sale (as was referred to previously). In this hierarchy, the obligation to give vacant possession (when appearing as a special condition) should take precedence over competing conditions. The decisions in *Curtis* and *Korogluyan* allowed general conditions to negate the express special condition for vacant possession, contrary to established principles. The decision in *Topfell* supports the precedence of special conditions in the context of the entire contract. Seeking to claim that *Curtis* and *Korogluyan* should be preferred would do violence to the purpose and effect of the status of general and special conditions and the certainty that the hierarchy creates in the context of such contracts.[45]

Second, given that the decision in *Topfell* gives substance to the integral obligation of a sale and purchase contract, namely that vacant possession will be given, preferring that decision would seem logical from a practitioner's perspective. In chapter 1, the importance for parties giving, and receiving, vacant possession was highlighted. Vacant possession is an essential, if not *the* essential part of a contract for the sale and purchase of land, as various commentaries have noted.[46] The importance, and 'assumption' as to vacant possession being given in such cases is what led to the obligation to be implied by the courts, as a matter of law, where there was no express condition for vacant possession in contracts of this kind.[47] If the decision in *Topfell* is not preferred, the essential element of such standard contracts

---

[43] Harpum, 'Vacant Possession - Chamaeleon or Chimaera?' (1988) *Conveyancer and Property Lawyer* 324, 400 (CH). Emphasis added.

[44] ibid at 400.

[45] The precedence of a special condition for vacant possession was recently reaffirmed in the case of *Weir* v *Area Estates Ltd* [2009] All ER (D) 189 (Dec).

[46] See Williams, TC 'Sale of Land with Vacant Possession' (1928) 114 *The Law Journal* 339 in which he described vacant possession as 'an integral part of the contract'.

[47] Furmston, MP *The Law of Contract* (London, Butterworths Law, 2nd revised ed, 2003) s 3.21.

would be subordinated, with buyers unable to require sellers to deliver the property to them in a state contemplated by the contract. If other conditions can be argued to modify that obligation, the standard form contract would no longer 'behave in the way that it should'.[48]

Third, in the conveyancing process, a special condition (or express statement in the particulars of sale) that required vacant possession to be given on completion, is something that the parties would have specifically considered before recording expressly in their agreement. A similar amount of consideration most likely will *not* have been given to the general conditions of the sale and purchase contract, which are incorporated without special attention to detail. It is therefore 'fairer' from an equity perspective for the parties to be held to an express promise over other conditions incorporated only by reference. The fact that a special condition for vacant possession should even override other conflicting *special* conditions can be seen to embody Templeman J's regard to the vacant possession clause as having precedence over *all* other terms of the contract, given the integral place of vacant possession in the contract. As such, the decision in *Topfell* makes sense in the real legal world and with respect to the nature of the contract that the parties are entering into.

Finally, adding weight to the claim that the decision in *Topfell* should be viewed as preferred, the decision has been used to support arguments made in a number of subsequent decisions where the meaning of vacant possession was in issue.[49] By contrast, the decisions in *Curtis* and *Korogluyan* have never been approved of, or followed.

As such, there are a number of justifications, in a theoretical and practical sense, for supporting the decision in *Topfell* over previous judgments and thus providing certainty for the parties in cases where the contract contains an express special condition for vacant possession. It can therefore be understood that, subject to contrary indication, an express special condition for vacant possession will take precedence over other competing or contradictory contractual conditions (special or general).

Express (general conditions) for vacant possession

Whilst clarifying the substance of an express special condition for vacant possession, it is important to note that the decision in *Topfell* did not provide authority for all

---

[48] See *Shell UK* v *Lostock Garages* [1977] 1 All ER 481, per Lord Denning MR at 487; *El Awdi* v *BCCI* [1989] 1 All ER 242, per Hutchinson J at 253; *Bank of Nova Scotia* v *Hellenic Mutual War Risk Association (Bermuda) Ltd* [1989] 3 All ER 628; *Perry* v Sharon Development Ltd [1937] 4 All ER 390; *Lynch* v Thorne [1956] 1 WLR 303 and *Hancock* v BW Brazier (Anerley) Ltd [1966] 1 WLR 1317.

[49] See *Secretary of State for The Environment, Transport and The Regions* v *Baylis (Gloucester) Ltd* [2000] 3 PLR 61 and *E Johnson & Co (Barbados) Ltd Appellants and NSR Ltd Respondents [Appeal From The Court Of Appeal Of Barbados]* [1997] AC 400.

types of vacant possession obligations. This is because, in *Topfell*, the vacant possession term appeared expressly as a *special condition*. Indeed, all of the cases referred to above have in common the fact that the obligation to give vacant possession in each case was expressly a special condition. As chapter 4 will show, over time an express obligation to give vacant possession has sometimes appeared as a *general* condition (for example, the 1970 edition and 1973 revision of the *Law Society's General Conditions of Sale*, and the *Law Society's Standard Conditions of Sale* 1990).[50] It is unclear as to whether Templeman's decision would have changed in such circumstances.

In *Topfell*, the Court held that an express special condition as to vacant possession should take precedence over inconsistent conditions. This is logical because, as noted previously, in the conveyancing process a special condition (or express statement in the particulars of sale) that required vacant possession to be given on completion, will obviously be something that the parties will have specifically considered before recording expressly in their agreement. The parties should therefore be held to that promise over other (general) conditions incorporated by reference.[51] It is often the case that a similar amount of consideration will *not* have been given to the general conditions of the sale and purchase contract, which will have been incorporated without particular attention to detail. It is therefore arguable that a distinction should be made, or a different approach should be appropriate, where a *general condition* as to vacant possession contradicts another general condition (such as 'no annulment, no compensation' or 'subject to local authorities' requirements').

For example, the 1970 edition of the *Law Society's General Conditions of Sale* included a general condition for vacant possession as condition 3(1):

Unless the special conditions otherwise provide the property is sold with vacant possession on completion.[52]

And a 'subject to local authorities' requirements clause as general condition 2(1):

...the property is sold subject —
(c) to all requirements, proposals or requests (whether or not subject to any confirmation) of any such authority.[53]

---

[50] *The Law Society's General Conditions of Sale* (1970 ed) (London, The Law Society, 1970); *The Law Society's Contract for Sale* (1973 revision) (London, The Law Society, 1973); and the *Standard Conditions of Sale* (London, The Law Society, 1st ed, 1990).

[51] A special condition for vacant possession should even override another conflicting *special* condition, according to Templeman J in *Topfell*.

[52] *The Law Society's General Conditions of Sale* (1970 ed) (London, The Law Society, 1970) condition 3(1).

[53] ibid condition 2(1).

It is unclear as to which general condition would have prevailed in a case where, for example, a notice to enter pursuant to a compulsory purchase order was served between exchange and completion. One view would be that the reference to 'unless the *Special* Conditions otherwise provide the property is sold with vacant possession...' implies that other general conditions cannot modify the general condition as to vacant possession. However, this interpretation would not be possible in the 1990 edition of the *Standard Conditions of Sale*, in which the general condition as to vacant possession simply provided that:

> *The buyer is to be given vacant possession of all the property on completion;*
> *this does not apply to any part of it included in a lease or tenancy ('tenancy')*
> *subject to which the agreement states the property is sold.*[54]

Here, there was no suggestion that this general condition should take precedence over any other general condition, with each apparently having the same status. As such, Templeman J's statement that an express undertaking to give vacant possession (appearing as a special condition) constitutes an overriding guarantee by the seller, and should take precedence over other conditions, is not necessarily applicable where the obligation to give vacant possession appears as a general condition (with, one would assume, the same status as other general conditions).

Templeman J did hold that a special condition for vacant possession should even override another conflicting special condition. On the basis that the vacant possession clause (when a special condition) should have precedence over *all* other terms given its integral place in the contract, by analogy it could be argued that a general condition for vacant possession should take precedence over a conflicting general condition on the same basis (i.e. that the vacant possession condition is integral and should be given precedence over other incorporated terms of the same status). Whilst there is no authority on such a position, this analysis is within the reasoning of the decision of Templeman J. This does, however, cause the position for parties to remain unclear and this uncertainty creates greater risk for contracting parties who cannot rely on the law to assist them in seeking to interpret their rights and responsibilities in an instance of this kind.

It is also unclear what the position would be if a 'no annulment, no compensation' or 'subject to local authorities' requirements clause was a special condition (with vacant possession only a general condition). Whilst the lack of express contemplation does not detract from the contractual status of general conditions in a legal contract for the sale and purchase of land, it can legitimately be taken into account when determining the relative weight that may be attached to such a term of the contract by the parties, as compared to other conditions which may have been more particularly

---

[54] *Standard Conditions of Sale* (1st ed, 1990) condition 3(1).

considered (i.e. the special conditions).[55] According to established principles previously referred to, the special condition should prevail on the basis that the general conditions apply so far as they are 'not inconsistent' with the special conditions. A general condition for vacant possession would therefore only be permitted to have effect in so far as it did not qualify the conflicting special condition.

It is difficult to find a basis on which the decision of Templeman J could have application, or be used by analogy, in this scenario, unless one could suggest that the precedence that Templeman J gave to a condition for vacant possession could cause a general condition for vacant possession to be treated differently to other general conditions of the contract, and be an exception to the established rule that special conditions must take precedence. There is no authority supporting an assertion that a general condition for vacant possession should be treated as taking precedence over a conflicting special condition of sale. This further causes uncertainty for the parties when an issue of this nature arises, with neither being certain as to which can rely on the law to support their position. With that said, it is considered most likely that the conflicting special condition would take precedence (unless there was a contrary indication in the document) over a general condition for vacant possession. The incorporation of vacant possession in the current editions of the *Standard Conditions of Sale* is discussed in chapter 4.

Implied vacant possession conditions

The above discussion was set in the context of *express* vacant possession clauses, which have historically been included in a contract for the sale or lease of land as general or special conditions. However, there is a further permutation to consider if a term for vacant possession is not included in the contract expressly as a general or special condition. In such a case, as noted previously, it has been established in case law that the term that vacant possession will be given is, in such cases, *implied*.

In *Cook* v *Taylor*,[56] it was held that where a contract is silent as to vacant possession, and silent as to any tenancy to which the property is subject, there is impliedly a contract that vacant possession will be given on completion. Similarly, in *Midland Bank Ltd* v *Farmpride Hatcheries Ltd*, it was said that:

> ...*prima facie a prospective vendor of property offers the property with vacant possession unless he otherwise states and that would ordinarily be implied in the contract of sale in the absence of stipulation to the contrary.*[57]

---

[55] See Harpum, 'Vacant Possession - Chamaeleon or Chimaera?' (1988) *Conveyancer and Property Lawyer* 324, 400.

[56] *Cook* v *Taylor* [1942] Ch 349 at 352.

[57] *Midland Bank Ltd* v *Farmpride Hatcheries Ltd* [1981] 2 EGLR 147, per Shaw LJ at 148.

In *Edgewater Developments Co* v *Bailey*, it was said that 'where nothing was said about possession it was often said that there was an implication that property was to be sold with vacant possession'.[58]

However, it has been established that the implied obligation will not arise if that would be inconsistent with an express provision of the contract. In *Rignall Developments Ltd* v *Halil*,[59] Millett J spoke of 'the obvious impossibility ... of implying a term inconsistent with an express term of the contract'. Indeed, it has been held in Australia that '[a]part from any *special* conditions of sale it is the duty of the vendor to give vacant possession'.[60] That is, the obligation will be implied by law subject to an express special condition to the contrary. Further, an implied obligation to give vacant possession may also be subject to contrary express *general* conditions of the contract as well (i.e. not just special conditions, as suggested by Harvey J in *Reynolds* v *Doyle*). Indeed, Oliver LJ in *Squarey* v *Harris-Smith* said that any general condition having been incorporated:

> *must be given its full status as a contractual term and cannot just be ignored because it is one of a number of printed conditions which the parties may well not actually have read.*[61]

Therefore, if any impediment to vacant possession is irremovable, and at the time of contracting the obstacle was not known of (and could not reasonably be deemed to have been known of),[62] or if the impediment is removable, both general and special conditions may properly affect the nature, scope and extent of an implied obligation to give vacant possession (which *will* encompass such an obstacle) given that such conditions constitute part of the sale and purchase contract.[63] As such, an implied obligation to give vacant possession must be interpreted with reference to *general* and *special* conditions of the contract, both of which are capable of modifying the implied obligation.

---

[58] *Edgewater Developments Co* v *Bailey* [1974] 118 Sol Jol 312, per Lord Denning RM at 313.

[59] *Rignall Developments Ltd* v *Halil* [1988] Ch 190, per Millett J at 200. This was in an alternative context but the statement is of general application.

[60] *Reynolds* v *Doyle* [1919] 19 SR (NSW) 108, per Harvey J at 110. Emphasis added.

[61] *Squarey* v *Harris-Smith* [1981] 42 P&CR 118, per Oliver LJ at 128.

[62] If the seller knew (or ought to have known) of an impediment and had failed to make full and frank disclosure of it to the purchaser, then the seller will not be able to rely on any condition of sale in general terms which excludes or modifies its obligation to give vacant possession. The implied obligation to give vacant possession will not readily be excluded or modified in such a circumstance (see *Re Crosby's Contract* [1949] 1 All ER 830). This is an application of the rule that if there is any ambiguity, a condition of sale will be construed *against* the seller because it restricts the rights of the purchaser (see *Leominster Properties Ltd* v *Broadway Finance Ltd* [1981] 42 P&CR 372, per Slade J at 387).

[63] *Timmins* v *Moreland Street Property Co Ltd* [1958] Ch 110. If the irremovable defect *was* known of (i.e. was patent) then the implied obligation would *not* include such an obstacle — see chapter 2.

It is currently uncertain as to how other conditions (for example, 'no annulment, no compensation' or 'subject to local authorities' requirements clauses), which normally appear as general conditions, would be construed when there is only an implied obligation to give vacant possession. Whilst an express special condition for vacant possession has been held to override other conflicting (special and general) conditions, it does not follow that this should be the case where the vacant possession obligation is implied. In such a case, on the basis of established principles of the construction of documents, and in line with established case law,[64] it would be appropriate to give the conflicting express (special or general) condition(s) precedence over the implied vacant possession obligation on the basis that such a condition would constitute a 'contrary indication' or 'stipulation to the contrary'.[65] If this is correct, then any express condition having an effect on vacant possession would have to be given its full meaning *before* the implied obligation to give vacant possession could be given meaning.

Whilst there is currently no authority on this point, and the decision in *Topfell* does not have direct application to implied vacant possession terms, the decision in *Topfell* does reinforce the importance of a hierarchy of terms, and consistent with this would be to assert that expressly agreed terms are to be afforded greater weight than terms that are only implied, when a conflict arises. Again, the uncertainty surrounding this issue will cause difficulties for the parties to a contract who will be unsure of their legal rights and responsibilities. With that said, it is more likely that 'other conditions' of the contract (that may have an effect on vacant possession) would have to be given their full meaning above any contrary implied obligations. As such, there is a very strong argument for always making vacant possession an express term of the contract in any event.

Summary

The precedence of an express *special* condition for vacant possession over other conflicting terms was established by the decision in *Topfell* in 1979. This decision was clear that other special and general conditions could not be allowed to contradict the seller's contractual obligation to give vacant possession.[66] This decision supports the established hierarchy of special and general conditions and also makes sense in a practical context with respect to the implicit assumption that vacant possession will

---

[64] For example, *Rignall Developments* [1988] Ch 190, where Millett J at 200 spoke of 'the obvious impossibility ...of implying a term inconsistent with an express term of the contract'.

[65] In *Midland Bank Ltd* [1981] 2 EGLR 147, it was said that 'prima facie a prospective vendor of property offers the property with vacant possession unless he otherwise states and that would ordinarily be implied in the contract of sale in the absence of stipulation to the contrary'.

[66] *Topfell* [1979] 1 WLR 446, per Templeman J at 450.

be given on completion. The decision has also been approved of, and followed, in subsequent decisions, adding weight to the claim that it should be viewed as preferred. This assists parties to a transaction by enabling them to be able to assert their respective positions with reference to the understood and acknowledged precedence of an express special condition for vacant possession.

It was, however, noted that the obligation's interaction with other contractual terms, when appearing as an express *general* condition, was not as clear, and that *Topfell* could not be treated as an authority for all types of express condition for vacant possession. When the obligation to give vacant possession appears as an express *general* condition, competing or contradictory special conditions will logically take precedence given the established hierarchy of terms. It is arguable, however, that a general condition for vacant possession should take precedence over other conflicting *general* conditions on the basis of the reasoning (by analogy) in *Topfell*, namely that the vacant possession condition is integral and should therefore override other incorporated terms of the same status.

Further, the full nature and effect of an obligation to give vacant possession, when merely *implied* into the contract, also remains arguable and unclear. There is no authority on whether and if so, to what extent, special and general conditions of the contract should take precedence over an implied term for vacant possession and rebut the vacant possession implication established by case law, but logically express terms (whether special or general) would take effect over any implied terms.

The table below summaries the position in this regard:

Table 3.1 The Precedence of a Term for Vacant Possession and other Conflicting Conditions of the Contract

	Contradictory or Conflicting Special Conditions	Contradictory or Conflicting General Conditions
Special Condition for Vacant Possession	Subordinate to the special condition for Vacant Possession	Subordinate to the special condition for Vacant Possession
General Condition for Vacant Possession	Special conditions to be given full meaning first, in priority to the general condition for Vacant Possession.	Arguably of the same status — no authority on which would take precedence. *Topfell*, by analogy, implies that the Vacant Possession general condition would take precedence over other general conditions.
Implied Condition for Vacant Possession	As an express term, the special condition should take precedence.	As an express term, the general condition should take precedence, even if it is only one of many general conditions incorporated by reference and not particularly considered by the parties.

This demonstrates how fundamental questions relating to the term 'vacant possession', and its incorporation into standard contracts, remain unanswered at the present time, thus creating risk and uncertainty for the parties in any given instance. The current uncertainty that exists (in cases other than when vacant possession is an express special condition of the contract) causes difficulties to practitioners and those who are faced with the task of seeking to determine whether the obligation to give vacant possession is engaged in any particular instance, and what its full nature and effect will be in the context of the other conditions more generally. As discussed in chapter 1, whether an obligation has even arisen is one of the first questions that needs to be addressed. This can be difficult to ascertain if the obligation may be qualified by other conditions, or may not arise at all if other conditions rebut an implied obligation. Obviously this goes to reinforce why vacant possession should *always* be an express special condition of the contract.

The next chapter develops an analysis of vacant possession as a term of the contract, by explaining its current incorporation in the latest editions of the *Standard Conditions of Sale* and *Standard Commercial Property Conditions*. This further aids understanding of how a term for vacant possession can be incorporated into standard form contracts for the sale and purchase of land.

Chapter 4

Vacant Possession and Conditions of Sale

Chapter Outline

In chapter 3, the court's interpretation of terms for vacant possession, and their interaction with other contractual conditions, was analysed and discussed. The priority given to an express special condition for vacant possession over other conflicting terms was highlighted. This was contrasted with cases where there is a conflict with an express general condition providing for vacant possession, or an implied obligation for vacant possession; here the position remains less clear.

This chapter details a brief emergence and explanation of what are now commonly known as 'standard conditions of sale'. These were first published in 1902 and have since then been routinely incorporated into the majority of contracts for the sale and purchase of freehold land (and leasehold estates and interests) by reference. The chapter then explains the place of vacant possession in the current versions of the *Standard Conditions of Sale* (fourth edition) and the *Standard Commercial Property Conditions* (second edition), and the implications for transactions involving vacant possession. The chapter also highlights a number of particular issues relevant to practitioners when incorporating either set of the current conditions of sale into standard sale and purchase contracts.

▮ Origin of Conditions of Sale

Historically, precedent books published by eminent draftsmen have been a source of standard conveyancing provisions for use by practitioners across England and Wales. **67**

Vacant Possession: Law and Practice. ISBN: 978-0-08-096680-9

Thomas Martin and Charles Davidson, in *Practice of Conveyancing*, first published in 1837,[1] gave examples of such books from as far back as *The Chartuary* of 1534 and Dr Phayer's *Boke of Instruments* in 1543. Whilst early works dealt predominantly with techniques for the drafting of deeds, authors in time began to suggest standard contractual provisions. In 1790, in his *Original Precedents in Conveyancing*,[2] TW Williams included an agreement for the sale of a freehold estate where the purchaser was to be 'at the charge of the deeds' for conveying the property and 'all attested copies of the title deeds and covenants to produce the same'.[3]

In the second half of the nineteenth century, more complete sets of conditions were drafted in order to assist conveyancers in covering every aspect of the property transaction. Auction room conditions of the Liverpool Law Society, issued in 1865, are the earliest which texts have traced.[4] It was common practice for solicitors to prepare their own form of sale and purchase contract, including particulars and conditions. Indeed, various texts, including Farrer in 1902, set out general rules to be considered in 'framing conditions of sale and particulars'.[5] The emphasis in such texts was on how the individual solicitor should go about the bespoke drafting of the relevant particulars and sale contract unique to the specific transaction. Farrand notes that it was not until the early 1900s that:

> *practitioners apparently became more appreciative of their responsibilities when acting for the purchasers and began to contest any conditions [proposed by the seller's solicitors] going too far. As a result, sets of conditions of sale were drafted which could be applied to any sale and which endeavoured to adjust the balance more fairly between the vendor and the purchaser.*[6]

NATIONAL CONDITIONS OF SALE AND LAW SOCIETY CONDITIONS

The increasing professionalism of conveyancing and growth in the number of land transactions in the early 1900s is also said to have caused there to be a demand for a means by which transactions could be effected more smoothly and with greater

---

[1] Martin, T *Practice of Conveyancing, with Forms of Assurances* (London, Saunders and Denning, 1837). See also, Martin, T and Davidson, C *Practice of Conveyancing, with Forms of Assurances* (London, Saunders and Denning, 1837–1844); Stuart, J *Practice of Conveyancing* (London, Saunders and Denning, 1827–1831) and Stuart, J *Practice of Conveyancing* (London, Saunders and Denning, 2nd ed, 1832).

[2] Williams, TW *Original Precedents in Conveyancing* (Dublin, Zachariah Jackson, 1790).

[3] ibid at p 3.

[4] Wilkinson, HW *Standard Conditions of Sale of Land: a Commentary on the Law Society and National General Conditions of Sale of Land* (London, Longman, 4th ed, 1989) p 1.

[5] Farrer, FE *Precedents of Conditions of Sale of Real Estate, Revisions, Policies etc.* (London, Stevens and Sons Ltd, 1902).

[6] Farrand, JT *Contract and Conveyance* (London, Oyez Publications, 4th ed, 1983).

speed.[7] It was in 1902 that a solicitor's managing clerk from Norwich, Mr Alfred Kendall, suggested to the Solicitors' Law Stationery Society that a form of conditions of sale which he had prepared should be published. The draft was settled by Mr EP Wolstensholme and the first edition of the *National Conditions of Sale* was published around 1902.[8] Six editions were published under the pre-1926 law and 14 since commencement of the Law of Property Act 1925.[9]

In addition to the *National Conditions of Sale*, an alternative set of conveyancing precedents were published by the Law Society from 1926. Wilkinson[10] reported that the Law Society's Conditions were first published 'to facilitate conveyancing under the 1925 property legislation' and were used from 1 January 1926. They were intended to complement the Lord Chancellor's Statutory Conditions (which applied to contracts by correspondence)[11] and to relate to sales by public auction or by private agreement, other than by correspondence. They were first drafted by Sir Benjamin Cherry, who was previously involved in drafting the *National Conditions of Sale*.[12]

From 1925 there therefore existed two effectively 'competing' sets of conditions that could be incorporated into contracts for the sale and purchase of land: Oyez's *National Conditions of Sale* and the Law Society's *General Conditions of Sale*. Various editions and revisions of these competing sets of conditions were published up to 1990. No protocol was developed to govern which set of conditions should be used in any particular transaction and the forms of conditions could largely be selected on an ad hoc basis. Indeed, one commentator observed that:

Most solicitors habitually use one set of conditions in preference to the other, therefore some problems inevitably arise when a solicitor receives a purchase

---

[7] See (1926) 23 *Law Society's Gazette* 64 where facilitation of transactions, especially in light of the Law of Property Act 1925 taking effect on 1 January 1926, is discussed and Walford, EO *Conditions of Sale of Land* (London, Sweet and Maxwell Ltd, 1940) pp 2–3 with reference to the increase in conveyancing transactions.

[8] *National Conditions of Sale* (London, The Solicitors' Law Society Stationery Society Ltd, 1st ed, 1902).

[9] Amongst the draftsmen have been Mr T Cyprian Williams and Sir Benjamin Cherry (according to early versions of the *National Conditions of Sale*) and Walford, EO *Conditions of Sale of Land* (London, Sweet and Maxwell Ltd, 1940) p 2.

[10] Wilkinson, HW *Standard Conditions of Sale of Land: a Commentary on the Law Society and National General Conditions of Sale of Land* (London, Longman, 4th ed, 1989) p 1.

[11] The Statutory Form of Conditions of Sale, 1925 (better known as the Lord Chancellor's Conditions) applied to contracts for the sale of land made by correspondence. See Prideaux, F, Cherry, BL and Maxwell, JRP *Forms and Precedents in Conveyancing* (London, Stevens and Sons Ltd, 22nd ed, 1926) p 326. The Conditions were promulgated under s 46 of the Law of Property Act 1925 but applied only where no contrary intention was expressed. They are rudimentary and seem to have been little used (e.g. see *Stearn v Twitchell* [1985] 1 All ER 631).

[12] Wilkinson, *Standard Conditions of Sale of Land* (1989) p 1.

contract drafted on the set of conditions with which he is not familiar. Most problems arise quite simply from the fact that the solicitor is not fully aware of the differences between the two sets of conditions and therefore proceeds with the transaction on the assumption that the two sets contain more or less the same provisions.[13]

It was therefore apparent that two competing sets of conditions, drafted differently with mutually exclusive provisions, actually created the confusion and the very need for attention to detail in each particular case that these conditions, by replacing the previous bespoke drafting on a transaction specific basis, had sought to avoid.[14]

Standard Conditions of Sale

To facilitate everyday property transactions, and to prevent lawyers having to negotiate contracts on alternative sets of standard conditions, the first edition of the *Standard Conditions of Sale* was published by the Law Society in 1990. This was in conjunction with the launch, by the Law Society, of a Protocol for Domestic Conveyancing, which was an initiative which sought '...to bring about the standardization and simplification of conveyancing procedures for the benefit of the client'.[15] The Law Society's *Standard Conditions of Sale* were a fused version of the previously separate sets of conditions and superseded the two (previously separate) sets of conditions.[16] They were:

...intended to hold the balance evenly between the seller and the buyer and be entirely general in scope. They were radically different from their predecessors in arrangement and style and also make some significant changes in substance. [17]

The 1990 Conditions were broadly welcomed in the profession given that '...for the first time in living memory, practitioners may all be using one single set of

---

[13] Silverman, F *Conditions of Sale, a Conveyancers Guide* (London, Butterworth & Co (Publishers) Ltd, 1983) p 245.

[14] See Walford, EO *Conditions of Sale of Land* (London, Sweet and Maxwell Ltd, 1940) p 3.

[15] Silverman, F *Standard Conditions of Sale: a Conveyancers Guide* (London, Fourmat Publishing, 3rd ed, 1990) p v.

[16] The *Standard Conditions of Sale* (1st ed) 1990 were expressed to also be known and referred to as the *National Conditions of Sale* (21st ed) and the Law Society's *General Conditions of Sale* 1990.

[17] Silverman, F *The Law Society's Conveyancing Handbook 1993* (London, The Law Society, 1993).

conditions of sale in property transactions'.[18] In many respects, these fused conditions sought to achieve a situation originally contemplated in the early 1900s, where any sale could proceed on the basis of an established set of conditions which ensured a fair balance between the vendor and the purchaser (rather than alternative sets of different conditions, or bespoke conditions drafted on an ad hoc and transaction specific basis).[19]

Whilst originally the 1990 Conditions were not expressed to be confined to domestic conveyancing, but to be a 'total replacement'[20] to previous pre-fusion sets of conditions, from 1999 the Law Society's *Standard Commercial Property Conditions*[21] (based on the third edition of the *Standard Conditions of Sale*) were devised and these, co-existing with the *Standard Conditions of Sale*, are for use in specifically 'commercial' property transactions. As discussed below, the *Standard Conditions of Sale* (now appropriate mainly to residential conveyancing) are currently in their fourth edition, whilst the *Standard Commercial Property Conditions* are currently in their second edition.

Content of the Conditions of Sale

The Standard Conditions of Sale ultimately determine the parties' rights and obligations under the contract, and remedies in the event of a breach by either party. In terms of referring to the standard provisions as 'conditions', Cheshire[22] discussed the status of these conditions, and referred to Danckwerts LJ who commented on:

> ...*the longstanding practice which has arisen among conveyancers of referring to the provisions in a contract for the sale of land as 'conditions of sale' whether special or general (such as those provided by the common forms produced under the name of the National Conditions of Sale, or those produced by the Law Society)...*

Usefully, Danckwerts LJ went on to explain what was meant by the word 'condition' in this context:

> *The word 'condition' is traditional rather than appropriate, and these provisions are not so much concerned with the validity of the contract of sale as with the production of the title and the performance of the vendor's and*

---

[18] Silverman, F *Standard Conditions of Sale: a Conveyancers Guide* (London, Fourmat Publishing, 3rd ed, 1990) p v.

[19] As reported by Farrand, JT *Contract and Conveyance* (London, Oyez Publications, 4th ed, 1983) p 263.

[20] Silverman, *Standard Conditions of Sale* (1990) p v.

[21] *Standard Commercial Property Conditions* (London, The Law Society, 1st ed, 1999) and *Standard Commercial Property Conditions* (London, The Law Society, 2nd ed, 2003).

[22] Cheshire, GC and Burn, EH *Modern Law of Real Property* (London, Butterworths, 12th ed, 1976) p 74.

purchaser's obligations leading up to completion by conveyance. Shortly they are no more than terms of the contract.[23]

The various editions and versions of the conditions of sale over time have each included 'general' and 'special' conditions (or terms). In 1926, Davidson and Murray's guidance on conveyancing precedents noted that:

...it has become the practice...to embody in the contract a form of general conditions, which can be adapted or varied by special conditions applicable to the particular circumstances affecting the property to be sold.[24]

As the discussion in chapter 3 has noted, the general conditions deal with various issues relevant to the sale and purchase of property including, for example, insurance, deposits, requisitions and matters relevant to completion. They comprise a set of pervasive conditions that cover a variety of issues that *may* arise on a sale and purchase transaction, even though a great many of the conditions will never be relied upon in the majority of transactions. Special conditions, on the other hand, highlight specific aspects of the transaction that are especially of importance to the parties including, for example, interest rates and incumbrances on the property. These also provide an opportunity to address any unique factors relevant to the transaction that the general conditions do not adequately cater for. Cheshire[25] reported that a 'professionally drawn contract' will incorporate either set of standard conditions 'with variations to meet the particular case'. Matters which are commonly the subject of special conditions were, according to Cheshire, the root and length of title, the date on which possession was to be given, interest due (for example in the event that payment of the completion monies was delayed), mis-descriptions and relevant planning matters.

By agreement between the parties to the contract, the standard conditions of sale could be accepted in whole or in part and varied as required. Cheshire and Burn noted that:

The Law Society General Conditions of Sale and the National Conditions of Sale contain standard forms of conditions and these are usually employed with such alterations as the parties may make to fit the particular transaction.[26]

The conditions were, however, designed to facilitate common everyday property transactions by providing a universally recognised set of relevant conditions

---

[23] *Property and Bloodstock Ltd* v *Emerton* [1968] Ch 94, per Danckwerts LJ at 118.

[24] Davidson, C and Murray, AT *Concise Precedents in Conveyancing: with Practical Notes* (London, Sweet and Maxwell Ltd, 21st ed, 1926).

[25] Cheshire, GC and Burn, EH *Modern Law of Real Property* (London, Butterworths, 11th ed, 1976) p 741.

[26] Cheshire, GC and Burn, EH *Modern Law of Real Property* (London, Butterworths, 13th ed, 1982) p 110.

'intended to hold the balance evenly between the seller and the buyer...'.[27] The general intention was therefore that they should not be changed as this could make them more biased towards a seller or a buyer.[28]

The *Standard Conditions of Sale* themselves confirm that special conditions have priority over any inconsistent general provisions. The special conditions provide that the general conditions apply so far as they are not varied by or inconsistent with these special conditions. This was an established principle from the conception of the Conditions of Sale dating back to the early 1900s.[29]

Vacant possession in current Standard Conditions of Sale

Currently, lawyers and other professionals will routinely incorporate either the *Standard Conditions of Sale* (fourth edition) or the *Standard Commercial Property Conditions* (second edition)[30] into the contract.

STANDARD CONDITIONS OF SALE

As noted above, the *Standard Conditions of Sale* were first produced by the Law Society and The Solicitors' Law Stationery Society Ltd in 1990, and they superceded and replaced both the Law Society's Conditions of Sale and Oyez's National Conditions of Sale. The first edition of the *Standard Conditions of Sale* in 1990[31] altered the position on vacant possession compared to the previous respective sets of pre-fusion conditions, and dealt with vacant possession as a general condition under condition 3 marked 'Tenancies'. Condition 3.3.1 stated:

> *The buyer is to be given vacant possession of all the property on completion; this does not apply to any part of it included in a lease or tenancy ('tenancy') subject to which the agreement states the property is sold.[32]*

This, however, changed in the second edition:

> *Under the first edition, vacant possession was dealt with by the general conditions (the old condition 3.3.1). The current edition contains no provision relating to vacant possession, and accordingly the position must be dealt*

---

[27] *The Law Society's Conveyancing Handbook 1993* (London, The Law Society, 1993) p 665.

[28] Cheshire and Burn *Modern Law of Real Property* (13th ed, 1982) p 110.

[29] See commentary on the Law Society's Conditions of Sale 1953 at (1953) 97 *Sol Jol* 395.

[30] *Standard Conditions of Sale* (London, The Law Society, 4th ed, 2003) and *Standard Commercial Property Conditions* (2nd ed, 2003).

[31] *Standard Conditions of Sale* (London, The Law Society, 1st ed, 1990).

[32] ibid, condition 3.3.1.

with by a special condition (see the alternative versions of the printed special condition 4).[33]

In respect of the special condition provided for by the second edition, the alternatives to special condition 4 stated:

The property is sold with vacant possession on completion
OR
The property is sold subject to the following leases or tenancies.[34]

The third edition of the *Standard Conditions of Sale*[35] was published in 1995 and this replicated the second edition of the *Standard Conditions of Sale* with respect to the treatment of vacant possession. The fourth edition (2003)[36] also replicated the second edition and this edition remains the current edition used by conveyancers. As such, the position under the *Standard Conditions of Sale* at the present time is that vacant possession is a special condition in the form outlined above. It is left to the conveyancer to strike out the inappropriate alternative and, if the latter of the two remains, to list the leases and tenancies which the sale of the property will be subject to.

In seeking to make vacant possession a special condition, the explanatory note to the second edition highlighted how the special condition would override any 'inconsistent general condition' that was at odds with what the special condition now provided for. This reflected some awareness, for the first time, of the significance of the interaction between the obligation to give vacant possession and other contractual conditions, and the precedence of a special condition for vacant possession over other conditions of sale, as was established by the decision in *Topfell* v *Galley Properties Ltd.*[37]

A nuance between the drafting of the special conditions under the fourth edition of the *Standard Conditions of Sale*, and the *Standard Commercial Property Conditions* (second edition) is highlighted below.

STANDARD COMMERCIAL PROPERTY CONDITIONS

In 1999, the Law Society published a set of *Standard Commercial Property Conditions*[38] which were specifically intended to be used for medium to large commercial transactions (even though no definition of this was provided). These were based on the third edition of the *Standard Conditions of Sale* and were intended

---

[33] Aldridge, TM *Companion to the Standard Conditions of Sale* (London, Longman, 2nd ed, 1992) Pamphlet volume 115.

[34] *Standard Conditions of Sale* (London, The Law Society, 2nd ed, 1992) special condition 4.

[35] *Standard Conditions of Sale* (London, The Law Society, 3rd ed, 1995).

[36] *Standard Conditions of Sale* (London, The Law Society, 4th ed, 2003) special condition 4.

[37] *Topfell Ltd* v *Galley Properties Ltd* [1979] 1 WLR 446.

[38] *Standard Commercial Property Conditions* (1st ed, 1999).

to replace these in relation to commercial property transactions. From that point on, the *Standard Conditions of Sale* were primarily intended for use in residential and small commercial sales:

> *The SCPC [Standard Commercial Property Conditions] are intended primarily for use in more complex commercial transactions. Conveyancers are likely to find that, for residential sales and the sale of small business premises, the Standard Conditions of Sale (the 'SCS') are better suited to their needs.*[39]

Both the first edition (1999) and the second edition (2003)[40] of the *Standard Commercial Property Conditions* deal with vacant possession as a special condition. In the first edition of the *Standard Commercial Property Conditions*, special condition 3 replicated the special condition for vacant possession in the third edition of the *Standard Conditions of Sale*, namely:

> *The property is sold with vacant possession on completion*
> *OR*
> *The property is sold subject to the following leases or tenancies…* [41]

The second edition of the *Standard Commercial Property Conditions* made a small change to the wording of the special condition, namely:

> *The property is sold with vacant possession on completion*
> *OR*
> *The property is sold subject to* the leases or tenancies set out on the attached list but otherwise with vacant possession on completion.[42]

Whilst an estate being sold subject to a lease or tenancy or other such interest is the most common obstacle that would prevent the procurement of vacant possession on completion,[43] it is important to acknowledge that selling a property subject to a tenancy or other burden may, depending on construction of the document, give no guarantee of vacant possession *outside* the scope of that burden. This was something that appears to have been picked up in the second edition of the *Standard Commercial Property Conditions* which amended the special condition from the first

---

[39] *Explanatory Notes on the Standard Commercial Property Conditions* (London, The Law Society, 2nd ed, 2004) p 1.

[40] *Standard Commercial Property Conditions* (2nd ed, 2003).

[41] *Standard Commercial Property Conditions* (1st ed, 1999) special condition 3.

[42] *Standard Commercial Property Conditions* (2nd ed, 2003) special condition 3. Emphasis added. See also Abbey, R and Richards, R *A Practical Approach to Conveyancing* (Oxford, Oxford University Press, 9th ed, 2007) p 94.

[43] See chapter 6 for more detail.

edition to state that '[t]he property is sold subject to the leases or tenancies set out on the attached list *but otherwise with vacant possession on completion*'.[44] This changed from the first edition, which stated that '[t]he property is sold subject to the following leases or tenancies…'.[45]

NUANCE IN CURRENT CONDITIONS

The wording of the special condition to the second edition of the *Standard Commercial Property Conditions* clearly puts the seller under an obligation to give vacant possession of the property *other than* with respect to the part subject to the tenancy. No explanation is given in the explanatory note to the second edition as to why this change was made[46] but a possible reason for this subtle amendment, especially in the commercial context, is that a commercial or industrial property can often be large premises and a tenancy or lease can commonly exist with respect to only *part* of the premises. In the second edition of the *Standard Commercial Property Conditions*, the wording of the special condition causes an obligation to give vacant possession to arise *expressly* with respect to the *rest* of the premises not affected or subject to the lease, whereas in the old (first edition) of the *Standard Commercial Property Conditions*, no obligation with respect to the remainder of the premises would have arisen at all expressly. Whilst in the first edition of the *Standard Commercial Property Conditions* an *implied* obligation with respect to the rest of the premises may have been engaged (subject to the intention of the parties and general construction of the document),[47] the form of special condition in the second edition of the *Standard Commercial Property Conditions expressly* caters for vacant possession of the remainder of the premises in this way.

A comparison of the wording of the special condition for vacant possession in the *Standard Commercial Property Conditions* and *Standard Conditions of Sale* reveals that the third and (currently) fourth edition of the *Standard Conditions of Sale* (now used primarily for residential transactions) retains the alternative special condition from the second edition of the *Standard Conditions of Sale*, namely:

The property is sold with vacant possession on completion
OR
The property is sold subject to the following leases or tenancies.[48]

---

[44] *Standard Commercial Property Conditions* (2nd ed, 2003) special condition 3. Emphasis added.
[45] ibid.
[46] *Explanatory Notes on the Standard Commercial Property Conditions* (London, The Law Society, 2nd ed, 2004).
[47] *Rignall Developments Ltd v Halil* [1988] Ch. 190. This was in an alternative context but the statement is of general application.
[48] *Standard Conditions of Sale* (2nd ed, 1992) condition 4.

Use of this form of special condition for vacant possession is potentially an issue for sales of a partly tenanted residential dwelling, because vacant possession will not be given expressly with respect to the rest of the property (that is not subject to the disclosed tenancy) under this form of the special condition. This nuance between the current *Standard Conditions of Sale* (fourth edition) and *Standard Commercial Property Conditions* (second edition) is something of which practitioners may not be aware, and therefore they may not appreciate the full implications.

Whilst it is arguable that an *implied* obligation will be invoked (with respect to the rest of a given property) in connection with current use of the version of the special condition for vacant possession that appears in the *Standard Conditions of Sale* (fourth edition) (subject to contrary express provisions of the contract), practitioners are well advised to nevertheless consider amending the special condition provided in the *Standard Conditions of Sale* (fourth edition) to reflect the commercial counterpart in cases where:

- the alternative (b) is used (so that certain qualifications to the receipt of vacant possession are explicitly listed); and
- such qualifications (in the form of leases or tenancies) do not comprise the *entire* property which is the subject of the contract.

This will have the effect of ensuring that the contract includes an *express* obligation to give vacant possession in respect of the other areas of the property that are not subject to the disclosed lease(s).

Summary of vacant possession in Conditions of Sale

As noted in chapter 3, vacant possession will often be an express term of a contract for the sale of land. The decision in *Topfell*[49] established that an express provision (appearing as a special condition) that vacant possession will be given on completion will be the 'dominant' provision given that such a special condition is something that the parties will have specifically considered before recording expressly. General provisions of the contract should therefore be qualified by an express vacant possession clause (when appearing as a special condition) and not restrict the nature and effect of that clause in any way. Otherwise, a proper interpretation of the document's construction would derogate from the character of the express special condition for vacant possession, which amounts to a promise or guarantee that vacant possession will be given on completion.[50]

---

[49] *Topfell* [1979] 1 WLR 446, per Templeman J.

[50] As per Templeman J's decision in *Topfell* [1979] 1 WLR 446, as dicussed in chapter 3.

Under the *Standard Conditions of Sale* (fourth edition) and the *Standard Commercial Property Conditions* (second edition), vacant possession is currently a special condition, embodying the overriding guarantee and promise of vacant possession in standard-form contracts. With that said, a comparison of the two reveals that the (currently) second edition of the *Standard Commercial Property Conditions* has subtly amended the wording of the special condition for vacant possession in order to ensure that the obligation will still be engaged with respect to the remainder of the property that is not subject to a disclosed tenancy. This change was not fully explained upon implementation, and represents a nuance between the current (second) edition of the *Standard Commercial Property Conditions* and the current (fourth) edition of the *Standard Conditions of Sale*, the latter *retaining* the 'old style' alternative form of special condition for vacant possession.

Chapter 5

Defining the Obligation

Chapters 3 and 4 explained vacant possession as a 'term' of a sale and purchase contract which, under the current editions of the *Standard Conditions of Sale* and *Standard Commercial Property Conditions*, will be incorporated (by reference) as a special condition. This chapter explains the nature and meaning of the vacant possession term itself in more detail.

As explained below, the obligation to give vacant possession is shown to comprise two elements. First, a legal element, given that the obligation necessarily concerns a 'right to possession' which is transferred with the legal estate in land, but second, a factual element, that being 'actual possession' on completion, pursuant to the right to possession which is transferred. This factual element is further also shown to have a fixed time associated with it.

This chapter also contrasts the obligation with notions of constructive possession (i.e. possession *otherwise* than by actual occupation) in order to demonstrate why the legal and factual elements are intrinsic to the very essence of the obligation. This is essential to understanding how the obligation can be breached, which is the subject of chapter 6.

Interpreting the obligation

It is a fact that the term 'vacant possession' has never been authoritatively defined, with the area lacking judicial comment and debate: '[t]his [is] an area deficient in

79

Vacant Possession: Law and Practice. ISBN: 978-0-08-096680-9

legal authority…'.[1] Various leading counsel have struggled, in vain, to cite relevant case law to support the legal positions that they advance. In *Cumberland Consolidated Holdings Ltd* v *Ireland*, counsel for the claimant indicated that there was a lack of authority on which to base their submissions: '[t]hat does not assist in determining the meaning of "vacant possession" as between vendor and purchaser, a matter not decided by any authority'.[2] Judges have also made similar observations. Indeed, in the context of the meaning of the obligation to give vacant possession, it has been said that '… neither counsel was able to refer me to any authority which threw light on this problem'.[3]

Consequently, it is not surprising that judges have struggled with seeking to explain exactly what is meant by the term 'vacant possession'.[4] This has been especially the case given that the meaning of the words 'vacant possession' have been said to vary according to the context in which they are used:

> *[the] meaning of the words 'vacant possession' can, I think, vary from context to context.*[5]

In practice, the intention of the parties as shown by the contract will also be of importance.[6]

The statements above suggest that the meaning of the term 'vacant possession' has an inherently case specific element, making a universal definition difficult to arrive at. With that said, there would appear to be some fundamental legal principles associated with the obligation that can be extracted from case law in previous years and which are of universal application to the concept in whatever context it may arise.

The obligation to give vacant possession can be seen to have both a legal and factual dimension, and it is appropriate to explore the content of these constituent elements in more detail.

The legal element

The first element is the legal dimension to the obligation. Where a vendor expressly or impliedly contracts to convey an estate in land free from incumbrances, it has been established that it is, in principle, a term of the contract that the purchaser shall on completion be put into actual (and not constructive) possession.[7] It has been said

---

[1] *Sheikh* v *O'Connor* [1987] 2 EGLR 269.

[2] Submissions of Heilpern in *Cumberland Consolidated Holdings Ltd* v *Ireland* [1946] KB 264 at 268.

[3] *Sheikh* [1987] 2 EGLR 269, per Deputy Judge Wheeler at 274. See also Higgs, R 'Leave the Keys on your Way Out' (2005) 155 *New Law Journal* 149.

[4] See the submissions in *Cumberland* [1946] KB 264.

[5] *Topfell Ltd* v *Galley Properties Ltd* [1979] 1 EGLR 161, per Templeman J at 162.

[6] See *Lake* v *Dean* [1860] 28 Beav 607.

[7] See *Hughes* v *Jones* [1861] 3 De GF&J 307.

that: 'the phrase "vacant possession" is no doubt generally used in order to make it clear that what is being sold is not an interest in a reversion'.[8] This confirms that vacant possession is a constituent element of the legal transfer of the estate in land itself. This is the dimension of vacant possession which Deputy Judge Wheeler in *Sheikh* v *O'Connor* highlighted when he said that:

> [vacant possession] is a right, and it is a right which, in the absence of some competing legal claim, passes to the purchaser on completion [i.e. when the estate in land is transferred].[9]

This clearly showed that vacant possession was seen to be connected to the legal transfer of the estate in land, and associated rights which follow from such a legal transfer. In practice, however, it is clear that it is not confined to a legal transfer of the estate in land (and corresponding legal right to possession), but that the obligation also includes a factual element.

The factual element

The term 'vacant possession' can be seen to go beyond the legal transfer of the estate in land and rights consistent with the transfer, and be concerned also, on a practical level, with possession (in a factual sense) of the property in question:

> ...the right to actual unimpeded physical enjoyment *is comprised in the right to vacant possession.*[10]

Indeed, in *Cumberland*, Heilpern for the plaintiffs argued that:

> ...vacant possession is not limited in meaning to the absence of any adverse claim. This limited meaning only applies to cases relating to substituted service... That does not assist in determining the meaning of 'vacant possession' as between vendor and purchaser, a matter not decided by any authority. The right to vacant possession must give a right to physical enjoyment....Vacant possession must mean possession without impediments.[11]

These comments show that vacant possession is not just concerned with the legal transfer of the estate in land, and associated rights thereto, but that the obligation also concerns the exercise (in a factual sense) of the purchaser's legal rights with respect to the transferred estate. Indeed, vacant possession has been held *not* to have been

---

[8] *Cumberland* [1946] KB 264, per Lord Greene at 270.

[9] *Sheikh* [1987] 2 EGLR 269, per Deputy Judge Wheeler at 271.

[10] *Cumberland* [1946] KB 264, per Lord Greene at 272. Emphasis added.

[11] ibid, per Lord Greene at 272. Emphasis added.

given if the purchaser cannot actually enjoy the right of possession passed to him or her without first having to take legal action. This was a point specifically made by counsel in the decision in *Sheikh*:

> ...is 'vacant possession' given if the purchaser cannot enjoy the right of possession without first taking legal action?[12]

As such, the obligation to give vacant possession must be understood as more than just the legal transfer of the *right* to possession. Indeed, in *Royal Bristol Permanent Building Society v Bomash*,[13] Kekewich J made clear that transfer of the legal *right* to vacant possession is not all the purchaser contracts for. He said:

> *I do not think that a purchaser having a contract to sell with vacant posses-sion, is bound to take possession from the sheriff when he knows, as he did know in this case, that the man to be evicted, the man who had been holding over, was still [physically] on the premises and would have to be turned out by force. I think the purchaser is, under those circumstances, entitled to say,* 'Exercise your rights; first turn the man out, and then give me vacant posses-sion'. *Therefore I think the vendors were in fault, that they had contracted to give vacant possession, that they were not prepared to give vacant possession at the time when the contract ought to have been completed....*[14]

Here it was quite clear that the seller should exercise his rights (to obtain possession) first, before being able to give vacant possession to the purchaser, rather than passing the right to possession to the purchaser on completion of the contract when the estate in land was transferred. The decision in *Engell v Finch*[15] also supports this position. In this case a breach of contract arose due to the defendants not having taken the necessary steps to secure possession pursuant to the agreement. It was held to be irrelevant as to whether the defendants had taken all action open to them if possession could not be given on completion. This therefore confirmed the requirement, in a practical sense, that factual possession (as well as legal possession) needed to be given and only when the occupier has been 'turned out' (to use the language of Kekewich J) could vacant possession be given.

The judgment of the County Court at first instance in *Cumberland* (affirmed by the Court of Appeal), also supports this position, where it was said that:

> the words [vacant possession] were not limited to mean only that the purchasers would be given immediate and actual possession without any

---

[12] *Sheikh* [1987] 2 EGLR 269 at 271. Submission of Mr Cogley.

[13] *Royal Bristol Permanent Building Society v Bomash* [1887] 35 Ch D 390.

[14] *Royal Bristol* [1887] 35 Ch D 390, per Kekewich J at 394. Emphasis added.

[15] *Engell v Finch* [1869] LR 4 QB 659.

adverse claim to possession by any person rightfully *claiming: they meant also that the purchaser would be given such substantial, actual, physical and empty possession as would allow him* to occupy and use the property purchased...[16]

All these decisions highlight the need for the party receiving vacant possession to be able to occupy the said property in a factual sense (over and above having the *legal* right to possession which follows from the transfer of the estate in land). Indeed, as discussed in chapter 2 (and in more detail in the next chapter), the tests that case law has developed to determine whether vacant possession has been given are objective in nature and are concerned with whether the purchaser (or party with the right to vacant possession on completion) can occupy without difficulty or objection. The courts are required to determine whether the physical (or legal) impediment substantially prevents or interferes with the enjoyment of a substantial part of the property. These embody the practical dimension of vacant possession as a factual, as well as legal, matter. The next section expands analysis of the factual element of the obligation by focusing on the timing of this factual requirement.

Timing of the Factual Element

When understanding the factual dimension of the obligation to give vacant possession, it is important to note further that the tests referred to are applied on 'completion' (or the operative date). The factual element of the obligation to give vacant possession therefore concerns one's ability (on a practical level) to *actually* enjoy the right to vacant possession *immediately* on completion.

As noted above, case law confirms that the right to vacant possession also requires the purchaser to be able (as a matter of fact) to actually enjoy that legal right to possession: it refers to the practical, physical and factual sense of being able to immediately occupy the property (as distinct from *just* the transfer of the legal *right* to enjoyment that follows from the transfer of the estate in land). A number of judgments emphasise that this factual right to possession is at the point of completion:

> *I have come to the conclusion that [the sellers] were contractually bound,* on completion, *to hand over the ground floor in a condition which would allow the plaintiffs* to occupy it. *It is quite plain that at the date of the contract and at* the date fixed for completion, *the vendors cannot do that...The vendors cannot occupy it themselves, they cannot sell it to somebody who wishes to purchase it in order to go and live there himself and they cannot let it.*[17]

---

[16] *Cumberland* [1946] KB 264, per Lord Greene at 266. Emphasis added.

[17] *Topfell* [1979] 1 EGLR 161, per Templeman J at 162. Emphasis added.

Indeed, the fact of occupation on completion was also referred to in *Cumberland*:

> *...a vendor who leaves property of his own on the premises cannot ... be said to give vacant possession since [this is] inconsistent with the right which the purchaser has* on completion *to undisturbed enjoyment...*[18]

Here, the Court of Appeal in a case which primarily dealt with leftover chattels on the premises on completion, held that a vendor who left goods of his own on the property sold by him to an extent which deprived the purchaser of physical enjoyment of the property on completion, failed to give vacant possession. This reflected an awareness that the operative time for enjoyment was completion and a lack of enjoyment at that time would cause the obligation to have been breached: that is, the purchaser having received only the legal *right* to possession on completion would not be sufficient and would constitute a breach of the obligation. Other commentaries support the view that the purchaser must be able to enjoy the right to possession on completion:

> *Vacant possession, express or implied, requires the vendor to assign the property free from any claim of right to possession* ... and includes *'the right to actual un-impeached physical enjoyment' of the property ... the premises should be free* at completion *from any occupation by the vendor, a tenant, former tenant, a squatter and even material quantities of rubbish or furniture.*[19]

This further supports the analysis of the constituent parts of the obligation undertaken in this chapter, and in particular the timing of the factual element.

With respect to the day of completion itself, it is not general practice to stipulate that completion is required at a specific time on the day set for completion in standard sale and purchase contracts. The effect of this is to allow a seller to satisfy his or her obligation to give vacant possession if the purchaser secures possession at *some point* during that day. Chapter 9 addresses practical issues surrounding the giving of vacant possession on the day of completion itself in more detail, and how legal completion and the receipt of vacant possession must go hand in hand.

The elements and English land law

The legal and factual elements to the obligation to give vacant possession can be further understood with reference to English Land Law, and in particular the concept of 'possession'.

---

[18] *Cumberland* [1946] KB 264, per Lord Greene at 272. Emphasis added.

[19] Bacon, N *Conveyancing: Vendor's Duty of Disclosure* (Law Lectures for Practitioners, 1995) p 8. Available at http://sunzi1.lib.hku.hk/hkjo/view/14/1400190.pdf. Emphasis added.

The concept of 'possession' in English land law is fundamental and a number of sources emphasise how it is almost impossible to understand modern English land law without understanding the nature and significance of possession. It has been said that: '[t]hroughout the history of English land law the operative concept has been possession rather than ownership'.[20] Some have likened the importance of the concept of possession to other fundamental parts of the law of England and Wales, stating that '[p]ossession is a conception which is only less important than contract'.[21] Others have highlighted the predominant nature of possession in English land law. For example, Cheshire and Burn have noted that '[i]t has been said, rightly, that there is no law of ownership of land in England and Wales, only a law of possession'.[22]

In early law, possession was explained through the concept of 'seisin' which lay in the actual or *de facto* possession of land.[23] It is for that reason that, unlike ownership, which is seen as a *de jure* (legal) relationship between a person and a thing (and therefore a question of law),[24] possession is commonly viewed as a *de facto* (factual) relationship between a person and an object. Salmond wrote that 'whether a person has ownership depends on rules of law; whether a person has possession is a question that could only be answered as a matter of fact and without reference to law at all'.[25] It is true to say, however, that 'possession' can be a question of law as well as just fact alone. Indeed, the term possession can be used, and has been applied over time, in a number of distinguishable respects. It is essential to define these differing uses given that they have a direct bearing upon the differing meanings associated with the term. Explaining the meaning of possession from the literature surrounding it helps to interpret the obligation to give vacant possession more insightfully given that '[a]ny answer which does not distinguish between the different meanings of "possession" is inevitably going to be misleading at best, and simply wrong at worst'.[26]

LEGAL POSSESSION

Commonly, the term 'possession' is used to describe a relationship between a person (or legal entity) and an estate in land (for example, fee simple or a lease).[27] Legal

---

[20] Gray, K and Gray, S 'The Idea of Property' in S Bright and J Dewar (eds) *Land Law Themes and Perspectives* (Oxford, Oxford University Press, 1988) p 21.

[21] Oliver Wendell Holmes in Howe, M *The Common Law* (Boston, Little Brown & Co, 1963) p 163.

[22] Cheshire, G and Burn, E *Modern Law of Real Property* (London, Butterworths, 15th ed, 1994) p 26.

[23] See Lightwood, JM *Possession of Land* (London, Kessinger Publishing, 1894) pp 114–121; See also Hargreaves, AD 'Terminology and Title in Ejectment' (1940) 56 *LQR* 376.

[24] Wonnacott, M *Possession of Land* (Cambridge, Cambridge University Press, 2006) p 1.

[25] Salmond, JW *Jurisprudence* (London, Sweet & Maxwell, 12th ed, 1966) p 265.

[26] Wonnacott, M *Possession of Land* (Cambridge, Cambridge University Press, 2006) p 13.

[27] Here possession is not intended to describe the relationship between a person and any tangible property (such as a specific plot of land or a house).

possession, also referred to as *de jure* possession, signifies the 'right' to possession. Legal possession has been said to be enforceable in *rem* (that is, against the whole world at large), reflecting that such a proprietary right (to possession) is enforceable at law.

A person has a *right* to possess an estate if they have acquired a title to it which is 'vested in possession'. For a right to be vested in possession, the person or legal entity must have 'a present fixed right to it now'.[28] An example of an interest vested in possession would be the common scenario of the sale and purchase of a residential property where the transferor normally covenants to transfer the estate in land with the immediate right to possession for the purchaser on completion. This can be distinguished from a right to enjoyment[29] at some point in the future. Here the right to possession is vested only 'in interest'. An example of an estate only vested 'in interest' is a reversionary lease because it is granted to begin at some time in the future, usually after the prior existing lease has expired. A reversionary lease is 'vested' as soon as it is granted, but until the term begins it is vested only 'in interest' or 'reversion', and not 'in possession': it gives no present right to enjoy any estate in land.[30] The right to enjoy the estate is postponed to some future date, when its term will start.

The obligation to give vacant possession refers, in part, to the legal right to possession of the estate that has been transferred, normally pursuant to the sale and purchase contract. This is because the estate transferred is an estate 'vested in possession'. This is the first constituent element to vacant possession referred to earlier.

FACTUAL POSSESSION

The vernacular meaning of 'possession' is physical occupation of tangible land, also referred to in case law as 'actual' possession.[31] To have actual possession, a person must immediately have a sufficient degree of control over the thing in question.[32] Holmes wrote that 'a man must stand in a certain physical relation to the object and to the rest of the world, and must have a certain intent'.[33] The term 'actual

---

[28] Fearne, C *Contingent Remainders* (London, Stahan & Woodfall, 4th ed, 1844) Vol 1, p 2, cited with approval in *Pearson v IRC* [1981] AC 753, 772.

[29] That is, the exercise and use of the right and having the full benefit of it — see *Kenny v Preen* [1963] 1 QB 499, per Pearson LJ at 511.

[30] *Long v Tower Hamlets LBC* [1996] 2 All ER 683.

[31] *Prasad v Wolverhampton BC* [1983] 2 All ER 140.

[32] Panesar, S *General Principles of Property Law* (Essex, Longman, 2001) p 134.

[33] Holmes, OW *The Common Law* (Cambridge, Mass: Belknap, 1963) p 216.

possession' is sometimes used to denote the state of being 'in' possession of an estate, rather than merely having a right to possess it or having constructive possession of it. It has been said that 'in the modern law, *de facto* or actual possession is the closest to the ordinary or lay meaning of the term'.[34] The term 'natural possession' is also sometimes used along with occupation.[35]

With that said, a person who is physically present on land is in occupation of it. The presence might be personal, alternatively, the presence may be manifest through goods and chattels or agents or employees (for example, the estate may be enjoyed by or through third parties or actions such as the person leaving the property secured and unoccupied, or by putting into occupation a caretaker, janitor or security guard).[36] Further, a person who does not have a physical presence on land might, nonetheless, be treated as occupying it in certain cases.[37] An example would be so-called 'moveable fee simples', such as a tidal foreshore. Here, there is no *definite* area which the person in possession may occupy, but the person is still able to possess the fee simple by exercising the rights of the fee simple owner wherever the foreshore may actually be.[38]

As discussed above, the right to actual unimpeded physical enjoyment has been said to be comprised in the right to vacant possession,[39] a right that should be capable of immediate enjoyment on completion (or the operative date under a contract). It is therefore possible to associate this factual sense of possession as being the second constituent element of what the obligation to give vacant possession comprises. Indeed, it has been said that:

> ... *if an estate carries with it a right of occupation, then a person's possession of the estate is frequently made manifest by occupation.*[40]

This would seem a most appropriate means to explain the second constituent element of the obligation to give vacant possession, and the obligation more generally. As such, the second constituent element to vacant possession can be understood as the

---

[34] Wonnacott, M *Possession of Land* (Cambridge, Cambridge University Press, 2006) p 13.

[35] ibid at p 13.

[36] See *R v St Pancras Assessment Committee* [1877] 2 QBD 581, per Lush J at 588. This decision approved *Liverpool Corporation v Chorley Union Assessment Committee* [1912] 1 KB 270.

[37] *Bacchiochii v Academic Agency Ltd* [1998] 1 WLR 1313; *cf Esselte v Pearl Assurance* [1997] 1 WLR 981 and *Barnett v O'Sullivan* [1994] 1 WLR 1667. Wonnacott, M *Possession of Land* (Cambridge, Cambridge University Press, 2006) p 13 gives four examples of the dissociation between possession of an estate in the land and occupation of the physical land.

[38] See *Baxendale v Instow Parish Council* [1981] 2 All ER 620; *Jackson v Simons* [1923] 1 Ch 373 and s 61 of the Land Registration Act 2002.

[39] *Cumberland* [1946] KB 264, per Lord Greene at 272.

[40] Wonnacott, *Possession of Land* (2006) p 114.

fact of being 'in' possession or 'having' possession of the estate for which the right to possession has been passed. This, on a practical level, is also essential in the majority of cases (for example, where the purchaser is moving house and will want to take up occupation of the property on the day of completion, having moved from his or her current house on that day); that is, there is an inherently practical dimension to the receipt of vacant possession in any event.

CONSTRUCTIVE POSSESSION

A third meaning associated with the term possession is that of 'constructive possession'. This is commonly used in contrast to 'actual possession', so as to refer to possession of something *otherwise* than by actual occupation. 'Constructive possession' is also used commonly in a fictional sense to refer to the process by which the law deems a person presently to be 'in' possession of an estate, when, in fact, he or she is not.[41]

The right only to enjoyment at some point in the future would relate to a right vested only 'in interest' but not 'in possession'. A reversionary freehold is an estate only vested 'in interest' because it is granted to begin at some time in the future, namely after the prior existing lease has expired. A reversionary freehold gives no present right to enjoy any estate in land;[42] the right to enjoy the estate is postponed until some future date, as the tenant has the immediate right to possession vested in them. As such, the reversioner is only entitled to the receipt of rents and profits.

Commentators have used constructive possession to refer to the receipt of an estate in land subject to some prior interest (such as a lease or tenancy). Farrand explains that a sale subject to a pre-existing lease or tenancy would cause only constructive possession to be passed:

> *What the purchaser is entitled to get in the way of possession on completion depends, of course, on what the contract says. Thus, if the sale were made expressly subject to some tenancy or other, then the purchaser would only be entitled to constructive possession (i.e. the receipt of rents and profits)....[43]*

---

[41] ibid at p 114. The idea of constructive possession can be compared to the doctrine of 'constructive notice', which deems a party with having knowledge which they did not in fact have. See also Howell, K 'Notice: A Broad View and a Narrow View' (1996) *Conv 34*; Partington, D 'Implied Covenants for Title in Registered Freehold Land' (1988) *Conv 18* and Sheridan, D 'Notice and Registration' (1950) *NILQ 33*.

[42] *Long* v *Tower Hamlets LBC* [1996] 2 All ER 683.

[43] Farrand, JT *Contract and Conveyance* (London, Oyez Publications, 4th ed, 1964) p 259.

Yet, vacant possession is necessarily concerned with the entitlement to be put into 'actual possession' on completion or at the relevant time, pursuant to the right to possession which is passed with transfer of the estate in land. As Farrand continued:

> ...*if the sale were made expressly with vacant possession on completion, then the purchaser would be entitled to* actual possession, *i.e. in the sense of occupation...*[44]

Whilst possession is a term with varying meanings, in the context of vacant possession, it refers to both legal (*de jure*) and factual (*de facto*) possession. The legal possession manifest in the obligation relates to the passing of an estate in land that is vested in possession (i.e. not in interest or reversion). The factual element has been shown to relate to actual occupation of the estate transferred at the point of completion. Vacant possession is therefore not concerned with notions of constructive possession or deemed possession, by virtue of the receipt of rents and profits.

Summary of the elements of the obligation

The obligation to give vacant possession has two dimensions and necessarily concerns actual possession on completion pursuant to the right to vacant possession that has been transferred. The legal right to possession, and the factual ability to occupy, are two distinct, yet interrelated elements of the vacant possession obligation.

1. Vacant possession is the transfer of the estate in land (along with the corresponding legal *right* to possession thereto);
2. Vacant possession is only given when the party with the legal right to possession (which comes with the transfer of the estate in land) can:
   (a) *actually* enjoy that right of possession in a factual and practical sense
   (b) *immediately* at the point of completion (or at the operative time as provided for by the contract).

An appreciation of the obligation to give vacant possession as constituting both a legal and factual dimension is essential when considering how the obligation can be breached, arguably the most important issue for practitioners and those involved in transactions which include a vacant possession obligation. This is discussed further in the next chapter.

---

[44] ibid.

Chapter 6

Breaching the Obligation

Chapter Outline

The constituent elements of the obligation to give vacant possession were set out in chapter 5. The obligation to give vacant possession was shown to concern both the legal right to possession, and the ability for the right holder to actually possess the

91

Vacant Possession: Law and Practice. ISBN: 978-0-08-096680-9

land (in a factual sense). A review of literature defining key concepts such as possession and occupation assisted in interpreting the legal right to possession (*de jure*), and the factual ability to occupy (*de facto*), as two distinct yet interrelated elements of the vacant possession obligation.

This chapter focuses on what will amount to a breach of the obligation to give vacant possession, arguably the most important issue for practitioners and those involved in transactions which include a vacant possession obligation. Whilst the obligation must have an inherently legal and factual dimension, it is necessary to evaluate in more detail what will amount to a breach of the obligation at the point of completion. This is because whether or not there has been a breach of the obligation will be fundamental in determining the rights and corresponding responsibilities of the parties in question.

As referred to in chapters 1 and 2, there are three common types of impediment to the receipt of vacant possession. Whilst persons in occupation and legal impediments are often encountered, by far the most common obstacle will be items of a tangible nature being left in the premises at completion. This chapter looks at these three types of obstacle in turn, starting with tangible impediments.

Tangible impediments

The most common example of an impediment to vacant possession is when items that should have been removed by the seller (or party required to give vacant possession) are left at a property on completion. As noted in chapter 2, beer in the cellar,[1] furniture and goods remaining on the premises[2] and other chattels of the party required to give vacant possession[3] have been held to breach the obligation. In each case, the items being left at the premises were seen to be consistent with the seller keeping possession of the premises for his or her own purposes, and therefore inconsistent with passing vacant possession to the other party. As explained in *Scotland* v *Solomon*:[4]

> *a vendor who leaves property of his own on the premises cannot...be said to give vacant possession since [this is] inconsistent with the right which the purchaser has on completion to undisturbed enjoyment.*[5]

---

[1] *Savage* v *Dent* [1736] 2 Stra 1064.

[2] *Isaacs* v *Diamond* [1880] WN 75.

[3] *Cumberland Holdings Ltd* v *Ireland* [1946] KB 264.

[4] *Scotland* v *Solomon* [2002] EWHC 1886 (Ch).

[5] ibid per David Kitchin QC at 1887. This follows *Cumberland* [1946] KB 264.

Historic Position

Historically, decisions as to whether an obligation to give vacant possession had been breached generally proceeded on an ad hoc basis with respect to the particular case in issue. No general principles were established, either on the face of the cases or in conveyancing manuals or property law texts, to ensure consistency and continuity with respect to differing decisions on (ostensibly) similar facts. For example, in *Savage* v *Dent*,[6] leaving beer in a cellar was held to be consistent with the keeping of possession. As the summary to the case suggests:

> *The lessee of a publick house took another, and removed his goods and family, but left beer in the cellar. And there being rent in arrear, the landlord sealed a lease as on a vacant possession, delivered an ejectment, and signed judgment; which was set aside, the lessee still continuing in possession.*[7]

It was also noted that the same was true with respect to the 'leaving of hay' in a barn:

> *And a case was mentioned, where leaving hay in a barn at Hendon was held to be keeping possession....*[8]

No reference to the nature or quantity of the leftover goods was made. This was also the case in *Isaacs* v *Diamond*[9] where furniture and goods being left on the premises was held to be contrary to giving possession. Again, no analysis of the size or significance of these items with respect to the rest of the premises was undertaken. It was not until 1946, in *Cumberland Consolidated Holdings Ltd* v *Ireland*,[10] a case concerning rubbish that had been left at a property that was sold, that the Court first laid down what could be seen as a 'test' to determine whether vacant possession had been given; that is, a formulation for vacant possession that could then be reapplied in later cases.

Tests for Breach

In *Cumberland*, the plaintiffs contracted to buy a disused freehold warehouse from the defendants. By a special condition the property was sold 'with vacant possession on completion'. The cellars extending under the whole of the warehouse were made

---

[6] *Savage* [1736] 2 Stra 1064.

[7] ibid, case summary. Judgment not reported. In these cases, use of the term 'possession' (as opposed to vacant possession) featured prominently, although they can be interpreted as being synonymous.

[8] ibid.

[9] *Isaacs* [1880] WN 75.

[10] *Cumberland* [1946] KB 264.

unusable by rubbish, including many sacks of cement that had hardened thus making their removal particularly difficult. The defendant refused after completion to remove the rubbish and the plantiffs brought proceedings for damages for breach of the condition that the property would be sold with vacant possession on completion. The Court held that the defendant had failed to give vacant possession of the property sold. It was stated that a vendor who leaves his own chattels on property sold by him to an extent depriving the purchaser of the physical enjoyment of part of the property, failed to give vacant possession.[11] Such acts were consistent with the vendor seeking to continue to use the premises for his own purposes,[12] rather than passing possession to the purchaser in accordance with the terms of the contract. It was further noted that it was no answer for the vendor to claim to have abandoned his or her ownership of the chattels on completion to prevent a breach of the obligation. The Court held that a breach of the obligation would occur in cases where there was:

> ...the existence of a physical impediment, which substantially prevented or interfered with the enjoyment of the right of possession of a substantial part of the property, to which the purchaser did not expressly or impliedly consent to submit [i.e. agree to]...[13]

The *Cumberland* decision (as it shall hereafter be referred to), which remains the leading authority on this point, can be seen to have two distinct limbs. These limbs are explained in detail below.

Cumberland – Limb 1

The first limb of the test is directed at the activities of the party required to give vacant possession (i.e. tenant on exercising a break option in a lease, or seller when transferring an interest in land), and provides that if the conduct of the party in question indicates that the party, as seller or tenant, is continuing to use the premises for its own purposes in a non-trivial way (for example, by leaving goods in the premises), then the party will fail to establish that vacant possession has been given:

> *Subject to the rule de minimis a vendor who leaves property of his own on the premises on completion cannot, in our opinion, be said to give vacant posses-sion,* since by doing so he is claiming a right to use the premises for his own

---

[11] ibid at 268.

[12] See also *Norwich Union Life Insurance Society* v *Preston* [1957] 2 All ER 428 where a mortgagor that had left furniture in the premises after a court order requiring him to give up possession had not complied with the law and was using the premises for his own purposes as a place for the storage of his goods.

[13] *Cumberland* [1946] KB 264, per Lord Greene at 269. It is common for parties to agree that certain chattels may be left behind, and record this in the contract accordingly.

purposes, *namely, as a place of deposit for his own goods inconsistent with the right which the purchaser has on completion to undisturbed enjoyment.*[14]

As such, this first limb focuses on the specific circumstances of the party required to give vacant possession and its conduct in so purporting to give vacant possession. This limb can therefore be seen to be inherently fact specific in nature and to refer to the actual seller or tenant in question, and the party's intentions, belief and state of mind as manifest by its conduct.

Cumberland – Limb 2

The second test is directed at whether the contents of the premises present, objectively speaking, a substantial obstacle to the buyer's or landlord's own physical enjoyment of the premises on completion (or at the operative date when the obligation to give vacant possession is engaged). If they do, vacant possession will not have been given:

> *A vendor who leaves chattels of his own on property sold by him to an extent depriving the purchaser of the physical enjoyment of part of the property has failed to give vacant possession.*[15]

This second limb of the test – as to whether the contents of the premises, objectively speaking, present a substantial obstacle to physical enjoyment of the premises – was further elaborated upon in recent years in the context of the procurement of vacant possession when exercising a contractual break option in a lease.

The John Laing decision

In *John Laing Construction Ltd* v *Amber Pass Ltd*,[16] the claimant was the tenant of commercial premises under a lease granted by the defendant's predecessor-in-title. A clause in the lease provided that the lease might be determined by, *inter alia*, the 'yielding-up of the entirety of the demised premises'. As discussed further in chapter 9, the obligation to yield-up has been held to include the return of possession to the landlord. The claimant sought a declaration that, pursuant to a notice given under the break clause, it had validly terminated the lease. That claim was contested by the defendant, who sought to counter-claim for declarations that the purported break notice was ineffective and the lease was still continuing.

---

[14] ibid, per Lord Greene at 269. Emphasis added.

[15] ibid. See case summary.

[16] *John Laing Construction Ltd* v *Amber Pass Ltd* [2004] 2 EGLR 128.

The defendant contended that the claimant had not 'yielded-up' the property, relying, *inter alia*, on the continued presence of security guards at the premises and the claimant's failure to hand back the keys to the premises. It was argued that these were inconsistent with providing vacant possession at the end of the term. This did not persuade the Court and the claim was allowed. On the facts of the case, it was held that the claimant had plainly and obviously manifested a desire to terminate the lease and was accordingly entitled to the declaratory relief sought. The continued presence of security guards at the premises and the tenant's failure to hand back the keys were held not to have prevented vacant possession being given. The Court held that the task of the Court was:

> to look objectively at what had occurred and determine whether a clear intention had been manifested by the person whose acts were said to have brought about a termination to effect such termination, and whether the landlord could, if it wanted to, occupy the premises without difficulty or objection.[17]

John Laing and the first limb of Cumberland

The decision in *John Laing* can be seen to support the first limb of the *Cumberland* test, referring to the intention of the party required to give vacant possession:

> ... to ... determine whether a clear intention had been manifested by the person whose acts were said to have brought about a termination to effect such termination...[18]

This would suggest that some intent on the part of the seller or tenant, to manifest its desire to give vacant possession (through its actions and conduct) to the buyer or landlord, is relevant. This intention to vacate is equivalent to the first limb of the *Cumberland* test where the acts of the party required to give vacant possession are evaluated in determining whether the party's actions and conduct are inconsistent with the giving of vacant possession (for example, because the party is purporting to continue to use the premises for its own purposes, to store its goods and chattels).

John Laing and the second limb of Cumberland

The decision in *John Laing* also concurs with and further developed the second limb of the *Cumberland* test, which focuses on whether there exists a substantial obstacle

---

[17] ibid, per Robert Hildyard QC at 131.
[18] *John Laing* [2004] 2 EGLR 128, per Robert Hildyard QC at 131.

to the receipt of vacant possession on completion. With that said, some uncertainty regarding operation of the second limb of the *Cumberland* test has been apparent.

The problem with case law surrounding the second limb of the test is that it does not really help lawyers on a day-to-day basis when the facts of a particular circumstance have to be applied. For example, it is unclear what extent of difficulty is required and whether this must be general inconvenience or significant distress. *Cumberland* suggests that a seller or tenant has to remove all chattels and also rubbish which 'substantially prevents or interferes with enjoyment of a substantial part of the property', but there is no definition of what constitutes 'substantial'. It is also unclear whether there has to be an actual interference, or whether the likelihood or potential for the leftover items to cause a substantial interference will be sufficient.

The specific characteristics of the party with the obligation to give vacant possession (e.g. the seller or tenant) appear determinative in respect of the first limb of the test, which is directed at the party's own activities, and whether they are consistent with the party seeking to give vacant possession. If they leave behind chattels and other goods, such conduct can be seen to be suggestive of a lack of intention and commitment, on the seller or tenant's part, to vacate the premises. It can, however, be suggested that specific characteristics and contextual factors are also relevant to the so-called 'objective' second limb of the test in respect of the purchaser's commencement of possession/the landlord's resumption of use following the lease purportedly terminating. Indeed, it would appear that specific characteristics relating to the purchaser or landlord in question should be taken into account when determining whether the purchaser or landlord can (objectively speaking) (re)occupy without such an objection. Indeed, the so-called 'objective' second limb of the *Cumberland* test, as referred to in the *John Laing* decision, itself appears to refer to the specific characteristics of the party with the right to vacant possession:

> *and whether* the landlord could, if it wanted to, *occupy the premises without difficulty or objection.*[19]

As such, the second limb of the test is not completely objective, but is rather an objective test with reference to the particular circumstances and characteristics of the party with the right to vacant possession. The test should not therefore be judged against the average purchaser/landlord, but a purchaser/landlord with the particular qualities of the purchaser/landlord in question. This suggests that the Court should not consider more generally whether rubbish left at the property on the break of a lease or completion of a sale prevents the average purchaser or landlord (objectively speaking) from (re)occupying without difficulty or objection, but rather (objectively speaking) the *actual* purchaser or landlord in question given its specific characteristics and circumstances. As such, it is possible that a materially similar

---

[19] *Cumberland Holdings Ltd* v *Ireland* [1946] 1 All ER 284, per Lord Greene at 287. Emphasis added.

objection will be deemed valid in one context, but not in another, given the specific characteristics and contextual circumstances of the parties in question. As such, as far as the (objective) second limb of the test is concerned, whether or not vacant possession has been given will be a fact specific determination on a case-by-case basis with reference to the particular circumstances in issue.

Summary of tests

The decision in *John Laing* can therefore be seen to have reformulated, in slightly different words, the sentiment and spirit of both limbs of the *Cumberland* test and the focus on both the conduct and state of mind of the party required to give vacant possession, and whether objectively the premises in question are capable of occupation by the party with the right to vacant possession on completion, as the table below shows.

Table 6.1 A Comparison between the Requirements in Cumberland and John Laing

	Cumberland	John Laing	Comment
Limb 1	Does the conduct of the party required to give vacant possession suggest that possession is being passed?	Do actions of the party required to give vacant possession manifest a desire to give/return possession?	Directed at the intention of the party required to give vacant possession, as demonstrated by their actions.
Limb 2	Does the impediment (objectively speaking) substantially prevent or interfere with possession of the property or a substantial part?	Is it possible (objectively speaking) for the party with the right to vacant possession to (re)occupy without difficulty or objection?	Focusing on whether the party with the right to vacant possession is actually able (objectively speaking) to occupy the property.

The following example of a scenario in which the tests would be applied helps to explain their application.

Imagine that a tenant has trouble paying the rent on a lease and decides to move to smaller premises. The tenant exercises a break option in its lease which is conditional on vacant possession being given on the break date. On the break date the premises are empty except for a room left partially filled with building materials which the tenant failed to remove in time for the break date under its lease. The tenant claims that vacant possession was given and that the lease has come to an end. The landlord claims the converse and argues the lease will now continue until its contractual expiry in another 10 years time. The answer may very well turn on the nature of the tenant's business. If the tenant is a builder, the landlord will have a strong argument for saying that the tenant is still using the premises beneficially, for the storage of goods for the purposes of its business and therefore that a clear intention has not been manifested by the tenant to effect a termination of the lease. As such, the tenant will fail on the first limb of the *Cumberland* test.[20] Alternatively, if the materials (for example, carpeting tiles) were brought onto the premises by the tenant, who runs an office from the premises, to repair the premises (in compliance with its repairing obligations, for example) but it did not complete this in time, different arguments may apply, and the tenant might succeed in establishing that the materials remaining in the premises on completion were not consistent with the tenant continuing to use the premises for its own purposes (the first limb), and (if not great in number) that the building materials did not constitute a substantial impediment to the landlord's resumption of its possession of the premises (the second limb).

What the landlord, as the party with the right to vacant possession, will use the premises for (compare a large industrial warehouse to a small corner shop) may also be key in determining whether the leftover items are a substantial impediment to its receipt of vacant possession. It therefore remains unclear, principally due to a lack of relevant case law, as to whether a breach of the obligation would be found to have occurred on these facts.

The recent decision in *Ibrend Estates BV v NYK Logistics (UK) Ltd* [20a] does however suggest that the 'objective' second limb is much harder to satisfy than the first limb (under which the claim in *Ibrend* was allowed), and that the second limb is only likely to be made out in 'exceptional circumstances'. *Ibrend* also highlights that *only* limb 1 *or* limb 2, and *not* both limbs, needs to be made out in any given case in order to succeed in establishing that vacant possession has not been given (i.e. the limbs constitute *separate* tests in their own right).

De Minimis

De minimis is a Latin expression relating to 'minimal things', normally in the phrases *de minimis non curat praetor* or *de minimis non curat lex*, meaning that the law is not interested in trivial matters or that 'the law does not care about very small matters'.[21] The expression has also been used to describe a constituent or component part of a wider transaction, where it is in itself insignificant or immaterial to the transaction as a whole, and will have no legal relevance or bearing on the end result. In a more formal legal sense it means something that is unworthy of the law's attention. In risk

---

[20] This can be compared with *Legal & General Assurance Society Ltd* v *Expeditors International (UK) Ltd* [2006] EWHC 1008 (Ch); [2006] L&TR 22, where the judge decided that vacant possession had not been given because the warehouse was still being used for the storage of 'a few pallets and parcels in a largely empty warehouse'. It was noted that such items remained useful to the tenant's business. See also Fetherstonhaugh, G 'Can Premises that are Left Half Empty or Half Full be Vacant?' (2008) *Estates Gazette* 34, which provides an example in similar terms to the above.

[20a] [2010] PLSCS 186 − June 2010.

[21] See Ehrlich, E *Amo, Amas, Amat and More* (New York, Harper Row, 1985) p 100. Literally it means that 'the law does not concern itself with trifles'.

assessment, for example, it refers to a level of risk that is too small to be concerned with; some refer to this as a 'virtually safe' level.[22]

Historically, it was unclear as to whether a *de minimis* threshold operated in determining whether the obligation to give vacant possession had been breached, and this may explain the differing decisions reached by respective judges on ostensibly similar questions of fact.[23] More recently, case law has suggested that the obligation is qualified as being subject to a *de minimis* rule.[24] Lord Greene in *Cumberland* stated:

> Subject to the rule de minimis *a vendor who leaves property of his own on the premises on completion cannot, in our opinion, be said to give vacant possession.*[25]

Courts have previously made use of the term *de minimis* in a number of respects. For example, courts have dismissed copyright infringement cases on the grounds that the alleged infringer's use of the copyrighted work (such as sampling) was so insignificant as to be *de minimis*.[26] Under HM Revenue & Customs services guidelines, the *de minimis* rule can also apply to any benefit, property or service provided to an employee that has so little value that reporting of it would be unreasonable or administratively impracticable; the use of a company photocopier to make a small number of copies for personal use might be one example.

In the context of vacant possession, *de minimis* can be seen to refer to small or insignificant obstacles to the receipt of vacant possession. Whilst case law has indicated that the vacant possession obligation will be subject to a *de minimis* rule,[27] how that operates in practice and what such a threshold refers to, however, remains unclear and has not been elaborated upon by the courts. For example, in *Legal & General Assurance Society Ltd* v *Expeditors International (UK) Ltd*,[28] a rubbish bin, a table, coffee mugs and a swivel chair left at the premises were considered unimportant and justified no further reference in the decision on the point. Clearly, by themselves, the items would not have prevented vacant possession from being given. Conversely, in *Cumberland*,[29] rubbish that filled two thirds of the warehouse cellars led the Court to

---

[22] See the National Library of Medicine Toxicology Glossary — Risk De minimis.

[23] For example, in the cases of *Savage* [1736] 2 Stra 1064 and *Isaacs* [1880] WN 75.

[24] *Cumberland* [1946] KB 264.

[25] ibid, per Lord Greene at 270. Emphasis added.

[26] See *Bridgeport Music, Inc* v *Dimension Films* 383 F.3d 390, 393 (6th Cir. 2004), a decision of the US Court of Appeal.

[27] Following *Cumberland* [1946] KB 264 where the obligation was stated as being subject to such a rule.

[28] *Legal & General Assurance Society Ltd* v *Expeditors International (UK) Ltd* [2006] EWHC 1008 (Ch); [2006] L&TR 22. First instance decision. Upheld on appeal see *Legal and General Assurance Society Ltd* v *Expeditors International (UK) Ltd* [2007] All ER (D) 166 (Jan).

[29] *Cumberland* [1946] KB 264.

hold that vacant possession had not been given. It is difficult to draw the line when the facts lie somewhere between these two examples. What is clear, from the decision in *Cumberland*, is that the interference must be of some substantial nature:

> *When we speak of a physical impediment we do not mean that any physical impediment will do. It must be an impediment which substantially prevents or interferes with the enjoyment of the right of possession of a substantial part of the property.*[30]

As such, with respect to both limbs of the *Cumberland* test, it is clear that an element of substantiality is manifest in the determination as to whether the party seeking to give vacant possession is continuing to use the premises for its own purposes, and whether the party with the right to vacant possession can (re)commence occupation without difficulty or objection.

It would also seem that the quantity of items left, their size, movability and degree and purpose of annexation (issues not traditionally considered by early case law)[31] may be relevant factors in determining whether the items left cause a breach of the obligation to give vacant possession. Indeed, case law suggests that it may be relevant to consider the location of items in, around or outside the property concerned. In *Hynes* v *Vaughan*,[32] a vendor left large amounts of rubbish (rotting vegetation, soil, timber, broken glass paint tins and rubble) in the garden which, it was claimed, prevented the transfer of vacant possession. The rubbish was held to be consistent with the character of the property sold and could not be said to substantially prevent or interfere with the enjoyment of the right of possession of a substantial part of the property, since it was outside in the garden:

> *The state of this property of which complaint is made was, in my view, reasonably in keeping with the character of the property. There has been no suggestion that it was not reasonably consistent with the state of the property at the date of the contract... The concrete blocks and the wooden frames are neatly stacked. Their presence does not, by any stretch of the imagination, constitute 'an impediment which substantially prevents or interferes with the enjoyment of the right of possession'.*[33]

As discussed in the next chapter, with reference to the state and condition of the property, a different decision may very well have been reached if the rubbish had been inside the premises concerned, raising issues as to the scope of the obligation in a particular given context. For now, this case is useful in demonstrating that it is

---

[30] ibid per Lord Greene at 287.
[31] See, for example, *Savage* [1736] 2 Stra 1064 and *Isaacs* [1880] WN 75.
[32] *Hynes* v *Vaughan* [1985] 50 P&CR 444.
[33] *Cumberland* [1946] 1 All ER 284, per Lord Greene at 289.

difficult to interpret where a *de minimis* level may be set. Given the fact sensitive nature of a determination as to whether there has been a breach of the obligation to give vacant possession, it is impossible to objectively define a threshold at which the *de minimis* rule should operate in this respect. Practitioners need to be aware of this state of affairs and the effect it may have on their clients in any given circumstance.

BARGAINING IN THE SHADOW OF THE LAW

What is apparent in each case in which a decision has to be made as to whether a breach has occurred is that the determination tends to be based on a judgment of the court using the evidence available to the judge, rather than the application of a clearly delineated formula. In many respects this creates a situation where parties in dispute over whether vacant possession has been given (prior to court proceedings or in the conduct of litigation prior to trial) are 'negotiating in the shadow of the law'. As Cooter, Marks and Mnookin explain:

> *Pre-trial bargaining may be described as a game played in the shadow of the law. There are two possible outcomes: settlement out of court through bargaining, and trial, which represents a bargaining breakdown. The Courts encourage private bargaining but stand ready to step from the shadows and resolve the dispute by coercion if the parties cannot agree.*[34]

The courts have provided some guidance which can be extracted by practitioners in seeking to determine whether an obligation to give vacant possession has been complied with, but have fallen short of prescribing an actual formula which can be applied. In this respect, they have (perhaps unintentionally, inadvertently or simply by default) retained for themselves a great amount of discretion in being able to make a decision that they consider just and equitable. For example, as noted previously, in *Legal & General,*[35] a rubbish bin, a table, coffee mugs and a swivel chair left at the premises were considered unimportant and merited no further reference in the decision on the point, whereas, by contrast, in *Cumberland,*[36] rubbish that filled two-thirds of the warehouse cellars led the Court to hold that vacant possession had not been given. It is difficult to determine whether a breach has occurred when the facts lie somewhere between these two examples and a lack of available case law provides

---

[34] Cooter, R, Marks, S and Mnookin, R 'Bargaining in the Shadow of the Law: a Testable Model of Strategic Behaviour' (1982) 11 *The Journal of Legal Studies* 225–251.

[35] *Legal & General* [2006] EWHC 1008 (Ch); [2006] L&TR 22. First instance decision. Upheld on appeal see *Legal and General* [2007] All ER (D) 166 (Jan).

[36] *Cumberland* [1946] KB 264.

that the judge in question will have a great amount of discretion in deciding which side of the line his or her decision should fall (i.e. whether or not there is a breach of the obligation).

Whilst this state of affairs may assist the court, and the particular judge, it does not help the parties in dispute and the uncertainty that remains causes their negotiations to be undertaken in the 'shadow of the law', where issues such as personal circumstances, financial resource and contingent or connected commitments may cause one party to achieve a better deal than the other. The uncertainty that remains prevalent over whether the obligation has been breached in a given situation (due to the lack of a clear formula for determination and inadequate previous case law) can be seen to disadvantage the weaker party. Further, it can potentially cause settlements to not reflect the actual legal position as to whether, as a matter of fact, vacant possession was or was not given. Obviously if clearer, and more specific, guidance was available, there would be less discretion for the individual judge but conversely more certainty for parties litigating on the point.

The current situation therefore has the effect (albeit unintentionally or inadvertently) of placing the risk of any given dispute on the weaker party who may not be the author or creator of the risk, and who may not be responsible for the adverse consequences in question. This places a corresponding responsibility on professional advisers in transactions involving vacant possession in these circumstances, given that the merits of a particular claim may not be the only issue to properly evaluate in advising a client on the strategy that they should adopt to deal with the vacant possession issue.

THE STATUS OF LEFTOVER GOODS ON THE PROPERTY

When a seller leaves chattels on the property breaching the obligation to give vacant possession (or otherwise if the items are *de minimis*), in law the buyer is deemed to be an 'involuntary bailee' of the chattels. This means that the buyer takes possession of the seller's (or bailor's) chattels without having paid anything (i.e. without having given consideration for them). The effect of being an involuntary bailee is that the buyer cannot simply dispose of the chattels, but instead must take reasonable steps to safeguard the goods on the premises. What is reasonable will depend on all the circumstances. Normally, subject to any express provision in the contract, the owner will have the right to collect the goods following completion. In the leasehold context, leases will often expressly deal with the right of a landlord to dispose of any chattels left on the property upon lease termination.

It is advisable for the party acting as involuntary bailee to take an inventory of the seller's or former tenant's goods (and photographs of the interior of the premises) as soon as possible upon taking possession. This has two uses, first to protect the buyer or landlord from any accusations from the previous owner or tenant of theft of the goods. Second, if the seller or tenant refuses to collect the goods the buyer or landlord can serve notice on the owner of the leftover goods (under the Torts (Interference with Goods) Act 1977) requiring the owner to collect the goods, failing which the landlord or buyer will be able to sell the goods at auction with good title and pay the owner the proceeds of sale, less any expenses. This enables a buyer or landlord to lawfully dispose of goods and chattels left on the property.

These rules apply to any and all goods left in a given premises, and it is unwise for a purchaser to take a view that the goods are of no intrinsic value and to dispose of them without following this procedure.

SUMMARY

This section has demonstrated that whether a tangible obstacle to the receipt of vacant possession will constitute a breach of the obligation is a question of fact in each particular instance. Any determination will proceed on a case-by-case basis, with so-called 'rules of thumb' developing that can be applied to later cases.

With respect to the tests to determine a breach of the obligation to give vacant possession, the first limb of the *Cumberland* test is directed at the activities of the party required to give vacant possession, and is therefore specific to whether the actions and conduct of the party giving vacant possession manifest an intention to vacate the premises. The 'objective' second limb of the *Cumberland* test, which is directed at whether the premises are capable of occupation on completion, has been shown not to be strictly objective in nature, but that, in order to interpret the 'objective' test, regard must be had to the specific circumstances of the parties in question and (as will be expounded in more detail in chapter 7) the nature of the land or property itself. The decision in *John Laing* can therefore be seen to further support the respective limbs of *Cumberland*.

Certainly, at present, parties can be seen to be negotiating in the 'shadow of the law' in determinations relating to a breach of vacant possession by a tangible impediment, given that there is insufficient guidance in case law to accurately assess, or predict, how a court may rule in any given particular case. This causes issues such as bargaining strength and financial resource to be more salient considerations in parties' decisions to litigate or settle disputes. Practitioners can only refer to previous case law in areas of this kind and seek to draw on cases by analogy.

Persons in occupation

Case law has generally tended to deal with persons in occupation on the basis of whether they are lawful or unlawful occupiers. Whilst the established legal position is shown to be the same, regardless of the lawfulness of their occupation, over time the position with respect to unlawful occupiers can be seen to have been somewhat confused.

LAWFUL OCCUPIERS

There is a wealth of case law confirming that the presence of an existing tenant or other legal occupier at the premises on completion will prevent the giving of vacant possession.[37] This is commonly because a lease is still continuing (for example, the occupier has contractual or statutory rights to remain in occupation of the property) or because other persons with a lawful right to occupation prevent the delivery of vacant possession on completion (such as, licensees who are in the property).[38] A number of cases discussed in this section illustrate this scenario. As noted below, such cases (or the decisions in these cases) did not, however, centre on the meaning of vacant possession but rather merely confirmed (somewhat crudely) that the obligation had been breached.

An example is the case of *Sharneyford Supplies Ltd* v *Edge*,[39] where the plaintiff purchased land from the defendant under a contract that expressly provided for vacant possession on completion. The occupants refused to vacate the land and claimed the benefit of a business tenancy within the statutory provisions laid down in the Landlord and Tenant Act 1954. It was held that the occupants had a legal right to remain in occupation and accordingly the defendant was liable for failing to give vacant possession at the material time, and the claimant was entitled to damages.

A further illustration is provided by *Cleadon Trust Ltd* v *Davis*.[40] Here, the parties agreed to the sale and purchase of certain land. The land in question was, at the material date, occupied by persons who had formerly been tenants, but whose tenancies had expired. The tenants had, however, stayed on with the consent of the landlords and so were licensees. Accordingly, it was held that it was not possible for the vendor to give vacant possession in accordance with the contract at the relevant time because of the continued presence of these persons, and damages were awarded.

In keeping with these decisions is the judgment in *Leek and Moorland Building Society* v *Clark*.[41] Here, a joint tenancy was purportedly surrendered but this was

---

[37] For example, *Sharneyford Supplies Ltd* v *Edge* [1987] Ch 305; *Cleadon Trust Ltd* v *Davis* [1940] Ch 940; *Leek and Moorland Building Society* v *Clark* [1952] 2 QB 788 and *Beard* v *Porter* [1948] 1 KB 321.

[38] For a discussion of the problems of so-called 'sitting tenants', see Stocker, J 'The Problem of the Protected Sitting Tenant' (1988) 85 *Law Society Gazette* 14.

[39] *Sharneyford* [1987] Ch 305.

[40] *Cleadon Trust* [1940] Ch 940.

[41] *Leek and Moorland* [1952] 2 QB 788.

undertaken by only one of the joint tenants. The Court held that the purported surrender was insufficient to terminate the joint tenancy[42] and the joint tenants' continuing rights to remain in occupation therefore prevented the delivery of vacant possession on completion:

> *By agreeing to sell...with vacant possession Mr Ellison was, it seems to us, agreeing that the tenancy would be surrendered on completion. If he had been the sole tenant, completion would itself presumably have effected a surrender... In fact, the tenancy was one in which he and his wife were joint lessees, and, as will be seen, he never had any authority from her to surrender or terminate that tenancy. Though he is, of course, bound by the agreement which he signed, he may not have realised its effect. The question is whether in these circumstances the joint tenancy has been surrendered or otherwise terminated.[43]*

The case confirmed that the sellers, who had contracted to give vacant possession on a sale of the property subject to the joint tenancy, were unable to deliver vacant possession in accordance with the contract because of the continued presence of the wife, as a joint tenant who was not joined in the purported surrender of the tenancy on completion.

The case of *Beard* v *Porter*[44] was another case concerning residential occupation. Here, the vendor had agreed to sell to the purchaser a dwelling-house which was occupied by a sitting tenant with rights to remain in occupation pursuant to the rent restriction acts.[45] In reliance on a representation from the tenant that he intended to leave, the vendor expressly agreed that the purchaser was to be given vacant possession on completion. The purchase was completed, but the tenant then refused to quit the house. Given the tenant's statutory protection, the vendor had no means of compelling the tenant to adhere to his expressed intention to vacate; he therefore remained a lawful occupier at all material times:

> *Since it was* of the essence of the matter *that vacant possession should be given, and the plaintiff only entered into the transaction on that footing,*

---

[42] The decision in *Re Viola's Indenture of Lease* [1909] 1 Ch 244 was approved of and followed. The case of *Re Viola* concerned the right of determination conferred on husband and wife as joint lessees at the end of three years in a lease. The notice given pursuant to the lease was given by the husband only, and its validity was disputed on this ground. It was held that where a lease contains a proviso enabling the 'lessees' to determine the lease by notice, a notice given by one of two lessees will not, in the absence of evidence of authority from the other lessee to give it or of circumstances from which the Court can infer such authority, be effectual to determine the lease.

[43] *Leek and Moorland* [1952] 2 QB 788, per Somervell JL at 792.

[44] *Beard* [1948] 1 KB 321.

[45] The specific acts were not referred to in the judgment.

one would have expected the contract to take the form, usual in such cases, that completion would take place when vacant possession was given, so that, should the defendant fail to implement this vital part of his promise, the plaintiff would be entitled to treat the contract as at an end and abandon a transaction which had ceased to be of use to him.[46]

The presence of the sitting tenant was a clear breach of the obligation.[47]

Further, if a tenant is protected by the applicable rent acts and is in occupation, it has been held that the tenant cannot be evicted even if he or she joins into a contract to deliver up possession of the property. In *Appleton* v *Aspin*,[48] the seller contracted to sell a property to the purchaser. The seller's mother lived in the house under an occupation agreement within the Rent Act 1977, but joined in the contract (even though not paid to do so) promising not to exercise any right of possession against the purchaser. The seller's mother later refused to vacate and the purchaser claimed specific performance of the contract which provided for vacant possession. It was held that the seller's mother was not required to leave pursuant to section 98(1) of the Rent Act 1977. As such, the seller would not be able to deliver vacant possession on completion.[49]

UNLAWFUL OCCUPIERS

What is apparent from all the decisions referred to above is that they dealt with purportedly 'lawful' claims to occupation of the property (i.e. because of a statutory or common law tenancy or licence being in place on completion). The courts have, over time, questioned whether this remains the case with respect to persons who may be in occupation with no *lawful* claim or right (for example, squatters or trespassers).

---

[46] *Beard* [1948]1 KB 321, per Evershed LJ at 322. Emphasis added.

[47] Wilkinson, HW *Standard Conditions of Sale of Land: a Commentary on the Law Society and National General Conditions of Sale of Land* (London, Longman, 4th ed, 1989) p 4 suggests that this principle is also applicable to business leases where a statutory protected tenant may have purported to agree to move out on the completion of a sale of the freehold interest but then later reneges. See also *Reynolds* v *Bannerman* [1922] 1 KB 719; *Watson* v *Saunders-Roe* [1947] KB 437, CA and *Carter* v *Green* [1950] 2 KB 76, CA in relation to protected rights of tenants, along with the Rent Act 1977 and the Housing Act 1988.

[48] *Appleton* v *Aspin* [1988] 4 EG 123.

[49] As discussed in chapter 9, if the obligation on the party (in a leasehold context) is only to 'yield-up' the premises (which includes the return of possession of the premises), rather than to give vacant possession expressly, then the 'yield-up' obligation will not encompass an occupier who has a statutory right to remain in possession of the property. The covenant to yield-up the property will be construed against the statutory background of the occupation of the tenant — see *Reynolds* v *Bannerman* [1922] 1 KB 719. This can be contrasted to the position where there is an express obligation to give vacant possession on completion.

Indeed, there has been conflicting *obiter dicta* with regard to whether the presence of people in *unlawful* occupation at the point of completion breaches the obligation to provide vacant possession.[50]

Some statements suggest that the obligation would be breached in this situation, apparently on the basis that it is the duty of the seller (as the person responsible for providing vacant possession) to ensure that trespassers are evicted. For example, in *Cumberland*[51] it was noted that a seller's duty extends to removing unlawful occupants on completion. The case itself concerned leftover goods at the premises, but the judge considered (*obiter*) that the existence of a trespasser could be equated with a physical impediment preventing the delivery of vacant possession:

> *We cannot see why the existence of a physical impediment to such enjoyment to which a purchaser does not expressly or impliedly consent to submit should stand in a different position to an impediment caused by the presence of a trespasser.*[52]

This decision clearly treated physical/tangible impediments in similar terms to persons in unlawful occupation with respect to a breach of the obligation to give vacant possession. It therefore made no distinction on the grounds of the lawfulness (or otherwise) of the persons in occupation.

Other *dicta*, however, suggests that a seller would *not* be in breach by virtue of there being a person in unlawful occupation of the property at completion. In *Sheikh v O'Connor*,[53] the vendor contracted to sell a property to the plaintiff. Most of the property was tenanted but the vendor expressly contracted to sell one of the rooms with vacant possession. After completion, the purchaser complained that the room which should have been vacant was in fact occupied by one of the tenants as a trespasser. The purchaser sued the vendor for damages for his failure to give vacant possession. One of the issues was purely factual and concerned whether the tenant had taken possession of the room before, or after, the completion date. Deputy Judge Wheeler concluded that it had been *after* completion, which was sufficient for the action to be dismissed in favour of the defendant. However, the judge went on to consider (*obiter*) the position in the event that his finding of fact was incorrect and the trespasser had been in unlawful occupation of the premises at the material time.

The judge accepted that a vendor who had contracted to give vacant possession did not fulfill his contractual obligation if, at the date fixed for completion, there was

---

[50] For a discussion of the problems caused by unlawful third parties being in occupation on completion, and preventing the delivery of vacant possession, see Jones, PV 'Squatting and Squatting' (1991) 141 *New Law Journal* 1543.

[51] *Cumberland* [1946] KB 264.

[52] ibid, at 268, per Lord Greene at 270.

[53] *Sheikh v O'Connor* [1987] 2 EGLR 269.

a third party who had a *legal* claim to possession, but he considered the position to be different in relation to a trespasser. In such a case he considered that it was for the purchaser to seek his remedy in the county court against the trespasser, given that the legal right to possession had passed to the purchaser on completion. The judge posed the following scenario:

> *Suppose that a vendor (V) contracts to sell property to a purchaser (P) with completion fixed for say, March 1, that at some time prior to completion squatters break in, unknown to V or P, that on March 1, without visiting the property, the parties complete, the balance of the purchase price is paid and keys are handed over, and then on the following day P visits the property and discovers that the squatters have been there for several days. Is P then entitled to claim rescission or damages on the ground that V has failed to give vacant possession?[54]*

The learned judge took the view that the answer should be in the negative. He continued:

> *At most, I am inclined to think, V's knowledge puts him under an obligation to act reasonably as circumstances permit in the light of that knowledge. But this would not ...extend to requiring him to take legal action to evict the squatters, though he might at least be wise to put P in the picture.[55]*

These *obiter* comments suggested that a vendor will not be liable, even if the vendor expressly contracted to give vacant possession, in the event that persons with no lawful claim prevented the delivery of vacant possession on completion. This would leave a purchaser with no remedy against the seller and no legal right to sue or seek specific performance of obligations in the contract. The effect of this decision would therefore be to negate the obligation being operative in the sale and purchase contract between the parties. As was noted in chapter 3, in a sale and purchase contract the obligation to give vacant possession is fundamental and it is an essential element of such a contract that the buyer will want to be able to take possession of the property.[56] Such a determination in *Sheikh* is therefore entirely inconsistent with the nature and effect of the obligation to give vacant possession as being, perhaps, the most crucial part of the contract and the very reason for which the contractual relationship between the parties, pursuant to the sale and purchase contract, was formed (i.e. so that they can take possession of the property).

---

[54] ibid, per Deputy Judge Wheeler at 271.

[55] ibid, per Deputy Judge Wheeler at 271.

[56] Williams, TC 'Sale of Land with Vacant Possession' (1928) 114 *The Law Journal* 339 in which he described vacant possession as 'an integral part of the contract'.

These *obiter* comments in *Sheikh*, suggesting that the trespassers would not cause there to be a breach of the obligation to give vacant possession in such cases, conflict with earlier established authority on the point.

In *Royal Bristol Permanent Building Society* v *Bomash*,[57] the purchaser agreed to buy two houses, vacant possession of which was to be given on completion. When the day fixed for completion arrived, the houses were occupied by someone who was holding over unlawfully. It was held that the vendor was in breach of his obligation to give vacant possession on completion notwithstanding that the person in occupation had no right to be in the premises:

> *I think the vendors were in fault, that they had contracted to give vacant possession, that they were not prepared to give vacant possession at the time when the contract ought to have been completed, and that in fact the purchaser could not have got within a reasonable time that vacant possession for which he had contracted; and to that extent he has obtained something less than that which he contracted to buy.[58]*

Damages to compensate were awarded by the Court.

Similarly, in *Engell* v *Finch*,[59] the defendants, mortgagees of a house with a power of sale, sold it by auction to the plaintiff, the particulars of sale stating that possession would be given on completion of the purchase:

> *The defendants contracted to sell and deliver* possession *on the completion of the purchase...*[60]

The plaintiff soon afterwards contracted for a resale. On investigation the title was satisfactory, but on the plaintiff requiring possession before completing the purchase, it appeared that the mortgagor was in possession and refused to give it up. The defendants were in a position to have ousted him by ejectment, but failed to do so and subsequently refused to complete the sale.[61] On this basis, the plaintiff brought an action for a breach of the contract of sale and it was held that a breach of contract arose: that is, the mortgagor's continued unlawful presence on the property constituted a breach of the obligation to give possession on completion pursuant to the contractual terms.

In *Herkanaidu* v *Lambeth London Borough Council*,[62] the council sold one of its properties at an auction with vacant possession. The claimant was the successful

---

[57] *Royal Bristol Permanent Building Society* v *Bomash* [1886-90] All ER Rep 283.

[58] ibid, per Kekewich J at 291.

[59] *Engell* v *Finch* [1869] LR 4 QB 659.

[60] ibid, per Kelly CB at 663. Emphasis added.

[61] ibid at 663.

[62] *Herkanaidu* v *Lambeth London Borough Council* [1999] All ER (D) 1420.

bidder for the property. There was no completion on the due date and four days later the claimant's solicitors raised for the first time the question of squatters. The defendant's officers inspected the property, found no squatters and considered that the allegation was a device to avoid completion. Accordingly, a notice to complete was served thereafter. As completion had not taken place, the defendant rescinded the contract and informed the claimant that the deposit was forfeited. The claimant brought an action against the defendant reclaiming the deposit, valuers' fees and legal costs of the abortive purchase on the grounds (*inter alia*) that the defendant had been unable to provide vacant possession of the property. The Master rejected the claim and the claimant appealed. It was noted on appeal that where a potential physical impediment (in this case, squatters) was discovered pre-completion, a breach of the obligation to provide vacant possession would occur if it was not remedied before completion. That is, squatters would breach an obligation to give vacant possession at the relevant time:

> ...*where there is a potential physical impediment discovered before completion (as here) a breach of the obligation to provide vacant possession would only occur if it is not remedied before completion. This would be the case if a vendor remained living in the property or had furniture there prior to completion. The obligation would be to give vacant possession on completion and whether in fact there had been a breach of this would only be apparent if the purchaser tendered the money and placed the vendor under the obligation to give vacant possession. Breach would only occur when the vendor failed to do so.*[63]

Commentaries have also confirmed the position with regard to unlawful occupiers breaching the obligation to give vacant possession. In *Williams on Vendor and Purchaser* it is suggested that 'where property is sold with vacant possession, the vendor has to satisfy a purchaser that there is *no adverse claimant* and no occupier of the premises...',[64] and the context of this paragraph suggests that the reference to 'adverse claimant' should be understood as relating to an unlawful as well as lawful occupier. Further, Megarry and Wade, when referring to the *Sheikh* decision, also state that 'the better view is that it is the duty of the vendor to evict trespassers'.[65] As such, it can be taken to be immaterial as to whether the third party occupier is lawful or unlawful as that does not change their factual presence in the premises on completion, thereby causing a breach of the obligation to give vacant possession.

---

[63] ibid, per Mr David Vaughan QC at 1429.

[64] *Williams on Vendor and Purchaser* (London, Sweet & Maxwell, 4th ed, 1936) p 201.

[65] Megarry, W and Wade, W *The Law of Real Property* (London, Sweet and Maxwell, 7th ed, 2008) p 672.

POSSESSION BEFORE COMPLETION

Allowing the purchaser into possession *before* completion may also have an effect on the operation of an obligation to give vacant possession. This is an area fraught with danger and which can be seen to catch many people unaware of the potential implications.

In *Sophisticated Developments* v *Steladean and Moschi*,[66] a contract for the sale of land contained a clause that the purchaser would, from the date of contract, 'be responsible for the day to day management of the property and would take the rents and profits'. Delay in completion occurred and the vendor served notice to complete but the purchaser argued that the vendor had repudiated the contract because there were trespassers in occupation of part of the property when the notice to complete was served/expired. It was held that it was an arguable point as to whether the vendor was able to complete at the material time,[67] even though the purchaser had had day-to-day management of the property since exchange. That is, the purchaser's neglect (in allowing trespassers to take possession) potentially did not prevent him arguing that the vendor was unable to give him vacant possession under the contract.

It has also been held that a third party taking possession before completion can result in rights being established. In *Abbey National Building Society* v *Cann*,[68] Mrs Cann's son secretly obtained a mortgage on the purchase of her house but kept most of the money and defaulted. Mr Cann had been allowed into occupation just before completion and it was argued that this created an overriding interest which had priority over the mortgage company. The claim failed on the ground of fraud, but Dillon LJ considered that the occupation would otherwise have constituted an overriding interest; this would have prevented the delivery of vacant possession to a third party. In *Lloyds Bank* v *Rossett*,[69] builders were allowed into a house to renovate it before completion (supervised regularly by the purchaser's wife) and this was held to give the wife an overriding interest against the lenders of whom the wife had not known. At the date of completion, which was also the date of the loan, the wife's interest had become established and had priority; this would prevent the passing of vacant possession to a third party in accordance with an earlier contract for sale.

---

[66] *Sophisticated Developments* v *Steladean and Moschi* Unreported, 1978 CLYB 347 (CA).

[67] For further information about the requirements of being in a position to serve Notice to Complete, see Bowes, C and Shaw, K. 'Can I Have my Money Back?' (2008) *Property Law Journal* 204.

[68] *Abbey National Building Society* v *Cann* [1991] AC 56.

[69] *Lloyds Bank* v *Rossett* [1988] 3 WRL 1301.

All these cases demonstrate how unintended third parties can become barriers to the procurement of vacant possession on completion.[70]

Currently, the *Standard Conditions of Sale* (4th edition)[71] deal with occupation by the buyer prior to completion as condition 5.2, making the buyer a licensee with an obligation to quit the property when the licence ends. When a licence takes effect, condition 5.2 disapplies condition 5.1 in relation to the seller having responsibility for transferring the property in the same physical state as it was at the date of contract. Importantly, this can (and arguably, should) be interpreted as *not* referring to the responsibility for ensuring that *all* potential barriers to the separate and distinct obligation to give vacant possession (which forms part of the contract as a special condition) are removed prior to completion, an obligation which seemingly remains with the seller.

The *Standard Commercial Property Conditions* (2nd edition)[72] make no express provision in relation to possession before completion, or the effect of possession before completion on the obligation to give vacant possession. Under the first edition,[73] condition 5.2 provided for the buyer to be let into occupation before completion (in similar terms to the provision in the *Standard Conditions of Sale* (4th edition)), but this was removed in the second edition. In any event, the old provisions did not encompass the responsibility for removal of all potential barriers to the separate and distinct obligation to give vacant possession (which forms part of the contract as a special condition) as a consequence of occupation being taken up prior to completion by the buyer.

As such, if a party is to be allowed to take possession before completion then parties are best advised to modify any vacant possession obligation *expressly* to provide that a purchaser cannot rely on its own neglect or default (as a consequence of possession before completion) in claiming that a seller is either unable to give vacant possession on completion, or unable to serve a valid Notice to Complete because of the impediment to vacant possession. Depending on the circumstances, it may be that the obligation to give vacant possession should be expressed to have effect as at the (earlier) date that possession is taken by the purchaser, so that satisfaction of the obligation is judged at the time that the purchaser actually takes factual possession.

These recommendations are applicable in both cases where the *Standard Conditions of Sale* (4th edition) and the *Standard Commercial Property Conditions* (2nd edition) are incorporated, given that condition 5.1 in the *Standard Conditions of*

---

[70] For a discussion of the problems caused by unlawful third parties being in occupation on completion, and preventing the delivery of vacant possession, see Jones, PV 'Squatting and Squatting' (1991) 141 *New Law Journal* 1543.

[71] *Standard Conditions of Sale* (London, The Law Society, 4th ed, 2003).

[72] *Standard Commercial Property Conditions* (London, The Law Society, 2nd ed, 2003).

[73] *Standard Commercial Property Conditions* (London, The Law Society, 1st ed, 1999).

Sale (4th edition) does not expressly deal with the issue of vacant possession or all potential barriers to the receipt of vacant possession. Its disapplication of condition 5.1 in relation to responsibility for transferring the property in the same physical state as it was at the date of contract, is not (it can be argued) extensive enough to cover issues of vacant possession. This underlies the need to ensure that the issue is expressly dealt with either in the contract or by a supplementary legal document or deed at the point when earlier possession is to be given. Indeed, in the explanatory note to the second edition of the *Standard Commercial Property Conditions*, it is stated that:

> *The old condition 5...entitled 'Pending Completion' has been radically amended by removal of the provisions relation to occupation of the property by the buyer [before completion]. The view was taken that in the commercial context such occupation would be upon terms specifically negotiated and the general provisions of clause 5.2 were unlikely to be used to a significant extent.* [74]

This affirms the need for a separate agreement or set of provisions to cater fully with the terms of any earlier possession by a buyer, including (most saliently) the issue of vacant possession. This applies regardless of whether the *Standard Conditions of Sale* (4th edition) or their commercial counterpart are used. In both cases, bespoke drafting is required in order to adequately cater for the issue of vacant possession.

SUMMARY

Trespassers and others in occupation with no lawful claim will constitute barriers to the receipt of vacant possession, in the same terms as persons who remain in occupation with a lawful claim (i.e. protected tenants). If a contract has provided (expressly or impliedly) for vacant possession (i.e. subject to no subsisting third party occupation or rights to possession), then the obligation will be breached at the relevant time by the presence of any such persons.

Parties should also ensure that the taking of possession before completion is allowed only on terms which clearly modify a subsisting obligation to give vacant possession. Otherwise, a buyer may seek to rely on its default or neglect in claiming that vacant possession is not given by the seller when completion arises (or that the seller is not ready, willing and able to serve a Notice to Complete) because of an impediment which the buyer allowed or tacitly permitted to take effect on the premises whilst it was in (earlier) possession, prior to completion.

---

[74] *Explanatory Notes on the Standard Commercial Property Conditions* (London, The Law Society, 2nd ed, 2004) p 3.

Legal obstacles

It is possible that a legal obstacle may prevent the delivery of vacant possession on completion. So-called legal obstacles do not relate to physical items or persons, but impediments of a 'legal' nature.[75a] As discussed in chapter 1, examples of legal obstacles to vacant possession include the transfer of a strip of land subject to dedication as a public highway,[75b] on the basis that the highway authority has the right to possession of the surface rather than the owner of the sub-soil. As such, vacant possession cannot be given on completion. Another example is that of a property (with an existing first floor tenancy) being sold with 'vacant possession of the ground floor' but with a Housing Act notice limiting occupation of the whole house to one household. In *Topfell Ltd* v *Galley Properties Ltd*,[76] the seller contracted to sell a property that was partly tenanted and partly vacant. The facilities provided on the premises were inadequate for the existing occupants. After contract but before completion, the local authority served a notice under the Housing Act 1985, limiting the number of persons who were permitted to occupy the premises until additional facilities were provided. As a result, the vendor was unable to give vacant possession of the untenanted part of the premises.

In the recent case of *Weir* v *Area Estates Ltd*,[77] the claimant had contracted in 2008 to purchase freehold property with vacant possession (having successfully bid for it at auction). The register of title to the freehold estate included an entry of a nine-year lease of the property granted in 2004. The lease had purportedly been surrendered by the tenant in 2006, but notice of the lease had not been removed from the register. In the sale contract, the seller expressly contracted to give vacant possession, and other terms stated that the lease, whilst still referred to on the register, had been determined by operation of law, and that the buyer would:

> accept the position and shall not be entitled to require any further proof of the determination.[78]

At the time of the purported surrender of the lease, a petition in bankruptcy had been presented against the tenant who was subsequently declared bankrupt on that petition. As such, the lease was held not to have been validly surrendered pursuant to section 284 of the Insolvency Act 1986, which renders void any disposition of property by a bankruptcy in the period beginning with the day upon which the

---

[75a] Shaw, K 'More to it than meets the eye' (2010) *Estates Gazette* 4.

[75b] *Secretary of State for the Environment* v *Baylis and Bennett* [2000] 80 P&CR 324.

[76] *Topfell Ltd* v *Galley Properties Ltd* [1979] 1 WLR 446.

[77] *Weir* v *Area Estates Ltd* [2009].

[78] See *Weir* [2009] All ER (D) 189 (Dec).

bankruptcy petition is presented at court and ending on the statutory vesting of the bankrupt's estate in the trustee in bankruptcy.

The Court granted summary judgment to the purchaser to rescind the contract, and dismissed the seller's counterclaim for damages (given it had subsequently sold the property to a third party for a lower price than had been bid by the purchaser at auction). As the seller had contracted to sell with vacant possession, the seller could not convey the property with vacant possession until the lease had been validly surrendered or disclaimed by the tenant's trustee in bankruptcy. As this had not taken place, and the lease was held to still be in existence at completion, the seller was accordingly in breach.[79]

Whilst these cases provide clear examples of legal impediments to vacant possession, other cases can be seen to have provided an inconsistent picture as to whether vacant possession can be, and is, given at the relevant time. Such cases concern orders to requisition a property or the service of notices of compulsory purchase. Decisions which, on the surface, appear inherently contradictory can, however, be explained when analysed on the basis of an interpretation of the obligation to give vacant possession as involving a factual and legal dimension, as has been proposed in chapter 5.

The following discussion demonstrates that where the acquiring authority had actually taken (factual) possession, or had the legal right to possession vested in them (legal possession) at the date fixed by the parties for completion, the vendor was held unable to give vacant possession: the legal right to possession and the factual ability to occupy pursuant to that right (the elements of the obligation to give vacant possession) no longer being vested in the vendor.

As will be seen, the decisions, set in war time and when the Government required land for a specific public purpose, also show that the courts will take into account the wider context of the relevant circumstances in interpreting the obligation to give vacant possession. This further confirms the context specific nature of the obligation which was highlighted in previous chapters.

REQUISITIONING OF PROPERTIES

A small collection of cases concern the government requisitioning of properties under provisions of the Defence (General) Regulations 1939; the common set of circumstances to these cases being that the parties had entered into written

---

[79] The decision appears to have turned on particular insolvency provisions and their interpretation, and may well be appealed on the basis that a surrender of the lease had been validly effected. The effect of the decision, however, is to correctly confirm that a subsisting legal estate or interest will be a legal barrier to the receipt of vacant possession, as expressly contracted for, on completion (this principle will be unchanged by an appeal decision).

agreements for sale and purchase of a property that became subject to a requisitioning notice before completion. In war time, it was necessary for properties to be requisitioned for certain purposes and obviously important that the Government was given vacant possession pursuant to the requisitioning notice so that the property could immediately be put to official use. Whilst historic, these cases are relevant (by analogy) to current requisitioning cases.

Some cases are clear that a requisitioning notice will *not* create an encumbrance on the land so as to prevent a seller from giving vacant possession to the purchaser. Conversely, other cases suggest that such requisitioning *will* prevent the seller from delivering vacant possession to the purchaser at the material time pursuant to the contract.

Not an incumbrance

One such case purportedly demonstrating that the service of a requisitioning notice will *not* amount to an incumbrance that will prevent vacant possession being given, is the decision in *Re Winslow Hall Estate Company* v *United Glass Bottle Manufacturers Ltd.*[80] Here, a contract for the sale of land had been entered into between the parties. Following the entering into of the contract, but before completion, notice was given on behalf of the Government to the purchasers that it intended to requisition the land under the Defence (General) Regulations. Regulation 51, made under the provisions of the Emergency Powers (Defence) Acts 1939 and 1940, provided that:

> *A competent authority, if it appears to that authority to be necessary or expedient so to do in the interests of the public safety, the defence of the realm or the efficient prosecution of the war, or for maintaining supplies and services essential to the life of the community, may take possession of any land, and may give such directions as appear to the competent authority to be necessary or expedient in connection with the taking of possession of that land.*[81]

The purchasers took out a vendor and purchaser summons under section 49 of the Law of Property Act 1925, seeking rescission of the contract. They asked for a declaration that the vendors were unable to show a good title to the premises or to perform their part of the contract because they had impliedly contracted, or were estopped from denying that they had contracted, to give vacant possession and could not do so as a result of the notice. Alternatively, it was claimed that the sellers had impliedly contracted to convey free from any incumbrance not mentioned in the contract and were now unable to do so.

---

[80] *Re Winslow Hall Estate Company* v *United Glass Bottle Manufacturers Ltd* [1941] Ch 503.
[81] Regulation 51 of the Emergency Powers (Defence) Acts 1939 and 1940.

The 'giving' of the requisition notice was held not to create an incumbrance on the land. In giving his judgment, Bennett J observed that, in the context of the regulation which he was considering, there was no requirement for the giving of a notice. He explained:

> *There is no provision in the Emergency Powers (Defence) Acts 1939 and 1940, and there is no provision in the regulations to which I have been referred which makes it incumbent upon the Office of Works, or upon any other competent authority, to exercise the powers which reg 51 gives them to give notice of their intention so to do to persons whose property they propose to take under the provisions of the regulation. It seems to me really a polite intimation on the part of the government that they propose to act, and it does not, in my judgment, create any greater incumbrance upon the land of the purchasers which it is proposed to take under this regulation. Anybody's land in Great Britain to-day is liable to be taken under the provisions of this regulation. In my judgment, it is not possible to hold that the notice of 25 January created an incumbrance so as to prevent the vendors from performing the contract into which they had entered.[82]*

Further, at the date that was set for completion, the Government had not actually taken possession of the land, and as such it was held that the sellers *were able* to give vacant possession to the purchaser (who would then lose the land when the Government later took possession, pursuant to the notice that was binding on the property). However, the Court held that:

> *I am not going to decide what the position of the parties would have been if* possession *had* been taken *before the date fixed for completion or before the vendors were in a position to complete, since it appears from the evidence that on February 3, 1941, the vendors were in a position to hand to the purchasers a properly executed conveyance and to give them vacant possession of the property which they had contracted to sell.[83]*

Clearly, therefore, the judge did not consider that service of the notice *itself* prevented the sellers from giving vacant possession on completion, and the fact that the Government had not taken possession before completion seemed material in providing that vacant possession could be given at the operative time as between vendor and purchaser. However, the judge seemed to suggest that if possession had been taken before completion, then different considerations would have applied (as elucidated below).

---

[82] *Re Winslow Hall Estate Company* [1941] Ch 503, per Bennett J at 506.

[83] ibid. Emphasis added.

Will be an incumbrance

The case of *Cook* v *Taylor*[84] dealt with the effect of a requisitioning notice under the same defence regulations. Simonds J reached the conclusion that, a notice having been served, the appropriate government authority had in fact 'entered into possession', because there had been what he described as the 'symbolic handing over of the keys of the property' in question.[85] He referred to the decision of Bennett J in *Re Winslow Hall Estates*, and drew a distinction between that case and the case before him, holding that on the date fixed for completion in this case the vendor was not in a position to complete because the 'parting with the keys of the property' was, as he put it, equivalent to symbolic delivery of the property to the requisitioning authority:

> *In the first place it was said that the requisition notice and what took place before…the date fixed for completion…did not preclude actual possession from being given. I do not take that view. It seems to me that, from the moment when the requisitioning authority served the notice* and took the keys from the vendor, *the vendor was not in a position to give vacant possession and was not in a position to allow the purchaser to enter on the property. It does not appear to me to be material whether it was before or after February 25 that occupation was actually taken by those persons who ultimately became the occupants.*[86]

The judge distinguished between the similar case of *Re Winslow Hall Estates* with respect to the taking of possession:

> *On that part of the case I refer to the decision of Bennett J. in* Re Winslow Hall Estates, *the facts of which were not very dissimilar but differ in one vital point… the government had not taken possession of the land. In the report there is no reference to any taking possession of the land until occupation was taken by the persons concerned… The vital difference between that case and this, as it appears to me, is that here, as I hold, on the date fixed for completion the vendor was not in a position to do that which he had contracted to do and give vacant possession to the purchaser, for he had already, pursuant to a proper requisition,* parted with the keys of the property, *which is equivalent to symbolical delivery of the property to the requisitioning authority. From that moment he could not give vacant possession to the purchaser.*[87]

---

[84] *Cook* v *Taylor* [1942] Ch 349.

[85] ibid at 352.

[86] ibid at 352.

[87] ibid at 352.

Clearly, the service of the requisition notice under regulation 51(1), followed by the handing over of the keys to the acquiring authority between the contract date and the completion date, was held to have deprived the seller of the ability to give vacant possession on the latter date; possession was no longer vested in the seller. On the date fixed for completion, the seller could not give vacant possession to the purchaser, for he had already passed possession to the requisitioning authority. From that moment it was not possible for him to give vacant possession to the purchaser. As such, the taking of possession before completion was a determinative issue in the reasoning of the Court.

Other authorities, however, suggest that from the moment when the requisition notice was *served* on the sellers, the sellers were not in a position to give vacant possession; that is, the notice *itself* prevented the giving of vacant possession.

A third case dealing with the same regulations is the Court of Appeal decision in *James Macara Ltd* v *Barclay*,[88] where the defendant agreed to sell certain property to the plaintiffs. Vacant possession was to be given on completion. Following exchange, but before completion, a government department, as the competent authority under the same Defence (General) Regulations 1939, served the defendant with a notice requisitioning the property. The defendant's solicitors sent a copy of the requisition notice to the plaintiffs; and the plaintiffs subsequently gave notice to the defendant that they rescinded the contract on the ground of the defendant's inability to give vacant possession.[89] The defendant disputed this and contended that the requisition notice did not, upon its true construction, amount to an exercise of the power to enter into possession under the Regulations, and, in fact, no actual entry had been made. Uthwatt J, giving the only judgment of the Court of Appeal, observed that actual entry on the land was not necessary for the due exercise of the power to take possession under the regulations, and stated:

> *What is required is that the immediate interest—an interest in possession—entitling the Crown to control of the land should be at the disposition of the Crown. …the power to take possession has been effectively exercised, although de facto possession has not been obtained. If actual entry be not necessary, there can, we think, be no doubt that the power is effectively exercised by notice which fairly brings to the mind of the person affected that the power is being exercised. A present intention stated to be exercised and communicated to the persons concerned is sufficient.*[90]

Whilst the cases of *Re Winslow Hall Estates* and *Cook* were referred to in argument, they were not discussed in the single judgment delivered by Uthwatt J. Clearly, on

---

[88] *James Macara Ltd* v *Barclay* [1945] KB 148.
[89] ibid at 149.
[90] ibid, per Uthwatt J at 154.

the construction of the regulation in question, the Court of Appeal came to the conclusion that whilst actual entry would no doubt be one way of establishing possession and effecting the right conveyed by the regulation, such an actual entry on the land was *not* necessary for the exercise of the power. This is because there was not any *particular* provision, so far as the regulation was concerned, which would determine the way in which the power to take possession might be exercised. The first instance decision was therefore affirmed.

This clearly established that the service of the notice *itself* conferred the right of possession on the Crown, meaning that vacant possession could not be given as between seller and purchaser. On this basis, the decision in *Re Winslow Hall Estates* must be seen as overruled because, whilst the judge did not consider that he needed to decide what the position of the parties would have been if possession *had* been taken before the date fixed for completion in *Re Winslow Hall Estates*, on the basis that the right to possession was transferred when the notice was served (and not when actual possession was later taken) he should have considered that very question, and ruled that vacant possession could not be given on completion (given that the Government could be deemed to have already 'entered into possession' at the earlier point in time when the notice was itself served).

It is apparent that differing judges' interpretations of the defence Regulations determined the decisions that were arrived at and, further, that this was based on the effect of the Regulations in transferring possession to the acquiring authority (as is discussed in more detail below). The judges in each of the three cases sought to address the issue of whether the acquiring authority had actually taken factual possession, or had the legal right to possession, at the date fixed by the parties for completion.

Purposive interpretation

A purposive interpretation of the Regulations, reflecting the (then) war time situation and need for the Government's efforts to be unhindered, may also be inferred from the cases. Whilst none of the decisions specifically dealt with the meaning of the term 'vacant possession', universal to all the decisions is the fact that the Government was at no time hindered in achieving its objectives by any of the decisions reached.

In *Re Winslow Hall Estates*,[91] it was held that the Government had not actually taken possession of the land at the completion date, and as such it was held that the sellers were able to give vacant possession: that is, the decision did not adversely affect the Governments' objectives. In *Cook*,[92] Simonds J reached the conclusion that vacant possession could not be given on completion. He drew an (arguably artificial) distinction between the case before him and the decision of Bennett J in

---

[91] *Re Winslow Hall Estates* [1941] Ch 503.
[92] *Cook* [1942] Ch 349.

Re Winslow Hall Estates, holding that on the date fixed for completion in this case the vendor was not in a position to complete because the parting with the keys of the property was, as he put it, equivalent to symbolic delivery of the property to the requisitioning authority. This therefore enabled the Government to take up occupation as was required. In *James Macara Ltd*,[93] Uthwatt J observed that actual entry on the land was not necessary for the due exercise of the power to take possession under the Regulations, reflecting, it would seem, the Government's need for an immediate interest in possession.[94] It was held that a 'present intention stated to be exercised and communicated to the persons concerned' was sufficient.[95]

These observations perhaps reflect a contemporaneous interpretation of the obligation in the context of the specific circumstances during the war time period and the overriding need for the Government to have possession pursuant to the Regulations. This is reinforced by the way that the decision of Bennett J in *Re Winslow Hall Estates* can be seen to have been overruled (or was simply wrong in the first place) following the Court of Appeal decision in *James Macara Ltd*. Obviously, the specific circumstances of any given case cannot be ignored when interpreting the obligation to give vacant possession, indeed as noted before: '[the] meaning of the words "vacant possession" can, I think, vary from context to context'.[96] The terms of the Regulations will also be material in any given case in respect of the powers conferred by the relevant legislation, and care must be taken to analyse these.

Legal and factual issues

All the decisions mentioned above clearly demonstrate that where the acquiring authority had actually taken (factual) possession, or had the (legal) right to possession, at the date fixed by the parties for completion, the seller was held *unable* to give vacant possession: the legal right to possession and the factual ability to occupy pursuant to that right (the essential elements of the obligation) no longer being vested in the seller. A clear articulation of the nature and form of the two constituent elements of vacant possession therefore assists in understanding what otherwise appear to be irreconcilable decisions.

In any case concerning a requisitioning notice, and the issue of vacant possession, careful consideration should be given to the effect of the relevant provision of the governing statute under which the notice is served, and the effect on the seller or relevant party, with reference to both the legal right to possession and the factual ability to occupy. The extent to which the vacant possession obligation will, or will

---

[93] *James Macara Ltd* [1945] KB 148.
[94] ibid, per Uthwatt J at 154.
[95] ibid.
[96] *Topfell* [1979] 1 EGLR 161 per Templeman J at 162.

not, be capable of being complied with will then be ascertainable in the given context more generally.

COMPULSORY PURCHASE ORDERS

Cases relating to compulsory purchase orders also appear, on the surface, to provide an inconsistent account of the meaning and context of the obligation to give vacant possession. However, when considered in detail, and with reference to the preceding analysis, such cases can be found to further assist in understanding the obligation to give vacant possession in its factual and legal senses.

A compulsory purchase order allows certain bodies which need to obtain land or property to do so without the consent of the owner. It may be used, for example, when developing infrastructure (e.g. new roads) where a land owner does not wish to dispose of the affected land. In respect of such orders, the authority acquiring the land or property may serve a 'notice to treat', which is an invitation (by the acquiring authority) to negotiate with the owner of the land that the authority wishes to procure. Once the acquiring authority has served the notice to treat, and if private negotiations are not successful, it is entitled to serve a 'notice of entry' which enables the authority to take possession of the land pursuant to the compulsory purchase order.

Where a compulsory purchase order is made over property between exchange and completion, one question that has arisen is whether the purchaser could claim that the contract has been 'frustrated' and that, as a result, the purchaser is not obliged to complete. In such cases, if frustration could be claimed, the obligation to give vacant possession would no longer arise as parties would be discharged from their obligations under the contract.

Frustration?

If a contract is made, and for whatever reason it later becomes impossible for one party to perform its obligations, then the doctrine of frustration may apply. The particular situation in question may have been expressly provided for in the contract, in the context of a *force majeure* clause. Alternatively, an event may take place that was not contemplated by the parties but which renders further performance impossible. Examples include the destruction of the subject matter of the contract,[97] the unavailability of an employee in an employment contract,[98] or a subsequent change in the law or circumstances which makes performance illegal.[99]

---

[97] *Taylor* v *Caldwell* [1863] 3 B&S 826.
[98] *Condor* v *The Barron Knights* [1966] 1 WLR 87 and *Hare* v *Murphy Bros* [1974] ICR 603.
[99] *Denny Mott & Dickson Ltd* v *James B Fraser & Co Ltd* [1944] AC 265; and *Ibrosa* v *Fairbairn* [1943] AC 32.

Most appropriate in the context of vacant possession is the unavailability of the subject matter of the contract (i.e. the property that is being compulsorily purchased).

In *Re Shipton, Anderson & Co*,[100] the owner of a specific parcel of wheat in a warehouse contracted to sell it on the terms 'payment cash within seven days against transfer order'. Before delivery and before the parcel passed to the buyer, the wheat was requisitioned by and delivered to His Majesty's Government under the powers of an Act passed before the date of the contract. It was held that delivery of the wheat by the seller to the buyer had been rendered impossible by the lawful requisition of the product by the Government. As such, the seller was excused from performance of the contract.[101]

When a frustrating event occurs, the contract is automatically discharged and the parties are excused from their future obligations. As no one party is at fault, neither party may claim damages for the other's non-performance.[102] It is for this reason that a party may seek to claim that a contract has been 'frustrated' and that, as a result, the obligation to give vacant possession is discharged.

In *Korogluyan* v *Matheou*,[103] the question to be decided was whether notices served pursuant to the provisions of section 11 of the Compulsory Purchase Act 1965, stating that the acquiring authority would be entering upon the land, meant that it was no longer possible for the seller to give vacant possession. It was held that although he was still in possession, the seller was unable to give vacant possession on completion in accordance with the contract.[104] Whitford J said:

> ...the word 'possession' should be considered in what might perhaps be described as its *popular* rather than its technical sense, *and that if one considers the position of a person buying a property of this kind and buying it upon this basis, that they are expecting to get vacant possession when the purchase is completed, it would really be distorting language to suggest that if it was being sold to them in circumstances where there had been a compulsory purchase order and a notice to treat and a notice to enter, they were in fact getting anything which* could sensibly be described *as vacant possession...Were it not for the fact that I think the defendant's case fails on special condition 9 and general condition 6, I would for my own part have come to*

---

[100] *Re Shipton, Anderson & Co* [1915] 3 KB 676.

[101] See also *Bank Line Ltd* v *Arthur Capel & Co* [1919] AC 435.

[102] The general rule is that the 'loss lies where it falls', so no claim can be made for the value of a partially completed contract. See *Appleby* v *Myers* [1867] LR 2 CP 651.

[103] *Korogluyan* v *Matheou* [1975] 30 P&CR 309.

[104] However, the purchaser was deprived of damages by certain conditions of sale.

the conclusion that in fact at the relevant time the plaintiff was not *in a position to sell with vacant possession, in the sense in which I think those words ought sensibly to be construed* in the context of the whole transaction.[105]

Whilst these comments were *obiter*, this decision clearly suggested that the service of the notice prevented the delivery of vacant possession as was contracted for. The language of the judgment suggested that the judge was seeking to apply a common sense analysis to the context of the case, to determine whether what 'could sensibly be described' as vacant possession could be given in such a case (even though no actual definition was provided).

However, these *obiter* comments can be shown to conflict with established authorities, such as *Hillingdon Estates Co* v *Stonefield Estates Ltd*.[106] In this case, the parties agreed to the sale and purchase of certain land. The completion of the transaction was delayed, *inter alia*, by the outbreak of war, and at a time when the contract was still uncompleted, the local county council made a compulsory purchase order affecting the whole of the property. Notices to treat under the order were served on the vendors and on the purchasers. The purchasers claimed that on the date of the service of the notices to treat, they were discharged from their contract to purchase the property, alleging that they had entered into the contract on the footing that they would be able to develop the land after completion.[107] This would not be possible if the property was compulsorily purchased. They therefore sought a declaration that they were discharged from liability under the contract and entitled to the return of the deposit paid plus interest. Vaisey J did not regard the service of the notice to treat as a frustrating event. He said:

I cannot hold that the contract here has been frustrated fundamentally, or indeed, at all. The purchasers in this case are certainly no worse off than they would have been if they had completed their contract in a period rather less than 12 years from the time when they agreed to complete it. Had they completed the contract without the delay of 12 years, quite clearly the compulsory purchase order would have affected them. However that may be, I have to consider the matter as I find it; and taking into consideration the long delay which has taken place, I still think that the contract, so far from being frustrated, can and should be carried out.[108]

---

[105] *Korogluyan* [1975] 30 P&CR 309, per Whitford J at 311. Emphasis added.
[106] *Hillingdon Estates Co* v *Stonefield Estates Ltd* [1952] Ch 627.
[107] ibid.
[108] ibid, per Vaisey J at 631.

The Court held that the purchasers were treated as owners in equity as soon as a binding contract was made. The service of a notice to treat did not affect the vendors given that their interest was to receive the purchase money; it followed that the risk of compulsory purchase properly fell on the purchasers, who were not entitled to rescind because of a future incumbrance. The incumbrance was therefore *not* a frustrating event as far as the contract was concerned. Vacant possession could be given in accordance with the contract, even though this was not the substantive issue in the case which came before the Court.

The same decision was reached in *E Johnson & Co (Barbados) Ltd v NSR Ltd*,[109] a decision of the Privy Council, in which the Court held that the publication of a notice under section 3 of the Land Acquisition Act (Barbados) 1894, warning that land under the contract of sale was likely to be required for Crown purposes, did not frustrate the contract. It was held that it was to be presumed, in the absence of specific provision to the contrary, that the purchaser had agreed to accept the normal risks incidental to land ownership as from the date of the contract, including the risk of interference with land-owning rights by the Crown. Their Lordships referred to *Re Hillington Estates* where Vaisey J, in the context of a notice to treat served by an acquiring authority after exchange of contracts but before completion, had remarked:

> *No doubt these departmental interferences and interventions do make a very great difference to ordinary life in this country, but that does not mean that, whenever such interference or intervention takes place, parties are discharged from bargains solemnly entered into between them. In my judgment, it is the duty of the parties, in such a case as this, to carry out their obligations; and I cannot see that there is in this case any reason at all for supposing that there is either an implied term of this contract that it should be frustrated in the event which has happened, or that there has been such a destruction of the fundamental and underlying circumstances on which the contract is based as to justify my saying that the contract did not exist, or ceased to exist at the date when the notice to treat was served...[110]*

Their Lordships considered that these observations were equally applicable to the position in this case (i.e. after the publication of the section 3 notice) and also referred to the case of *Amalgamated Investment & Property Co Ltd v John Walker & Sons Ltd*.[111] Here, a building was entered in the statutory list of buildings of special architectural or historical interest a few days after the date of a contract for its sale.

---

[109] *E Johnson & Co (Barbados) Ltd v NSR Ltd* [1997] AC 400.

[110] *Re Hillington Estates* [1952] Ch 627, per Vaisey J at 634.

[111] *Amalgamated Investment & Property Co Ltd v John Walker & Sons Ltd* [1977] 1 WLR 164.

The listing had the effect of dramatically reducing its market value.[112] The Court of Appeal held that the risk of a building being listed was one that every owner and purchaser must recognise they are subject to, with the result that the contract was not frustrated. The judges considered that a section 3 notice could not amount to a frustrating event on the same basis.[113] What is most salient in the decision is that, in holding that the notice did not amount to a frustrating event, it was held that the limited powers accruing to the Crown pursuant to the issue of a section 3 notice did not extend to a right to *immediate possession:*

> *The crucial question was whether the possession which [the vendor] had been in a position to give on the date for completion was something other than vacant possession within the meaning of the contract. If the acquiring authority had the right to* immediate possession *at the completion date, then the vendor could not then give vacant possession, but a section 3 notice did not give the Crown any such right. Johnsons had been in a position to give vacant possession on completion...*[114]

It was clear that the effect of the notice, in terms of whether it took from the seller the right to possession, was analysed. As the notice did not give the Crown an immediate right to possession, the notice did not prevent the owner of the land from being able to give vacant possession. Accordingly, the vendors were in a position to give vacant possession on completion and the purchasers were in breach of contract for refusing to complete.[115] Most saliently, the judgment clarified that the rights of entry and inspection conferred by section 3 of the relevant statute fell well short of the right of possession conferred by a requisition notice under regulation 51 of the Defence (General) Regulations 1939 (such as was found in the decisions in *Re Winslow Hall, Cook* and *James Macara*), or by notice of entry under section 11(1) of the Compulsory Purchase Act 1969. This was because both of these provisions entitled the relevant authority to dispossess the owner of the land. In this case, it was held that the purchasers remained free to enter onto the land and to use it for any purpose, that is, they retained the *right to possession* and ability to occupy pursuant to that right. As such, the obligation to give vacant possession was therefore held to be 'capable of being performed' (i.e. they could pass legal possession with the immediate ability for the purchaser to factually occupy — the essential elements of the obligation to give vacant possession, as set out in chapter 5).

---

[112] ibid at 164.
[113] ibid at 164.
[114] *Re Hillington Estates* [1952] Ch 627 per Vaisey J at 634.
[115] *Amalgamated Investment* [1977] 1 WLR 164.

Further, the effect of the notice did not render the procurement of vacant possession something 'radically different from that which was undertaken by the contract'.[116] The possession that could be transferred was the possession that was contracted for by the parties. The judge held that:

> ... *a threat of compulsory purchase, and publication of a section 3 notice ...*
> *does not radically alter the nature of the contract of sale. What it does is*
> *simply to increase the likelihood of an existing albeit remote risk becoming*
> *an eventuality.*[117]

This demonstrated an appreciation by the judge that the context of the transaction was relevant to the Court's interpretation of what the obligation to give vacant possession meant, although this was not explicitly discussed. The determinative point from this case was that the issuing of a section 3 notice did not extend to a right to immediate possession so as to prevent the owner of the land from being able to give vacant possession to a purchaser *on completion*. This was the crucial finding for the purposes of whether vacant possession could be given on completion. The fact that the property may ultimately have been compulsorily purchased (pursuant to the notice) *after* completion, was irrelevant to whether the seller could pass possession (in law) and the right to actually occupy pursuant to that right *on completion*. Barriers to vacant possession which occur after completion are not relevant, thus supporting the analysis of the timing element to the factual part of the obligation as proposed in the previous chapter.

Clarification of authority

The decision in *E Johnson & Co* casts doubt on the *obiter* comments of Whitford J in *Korogluyan*, which suggested that the service of a notice under section 11 of the Compulsory Purchase Act 1965 would prevent vacant possession being given, even though the seller remained in possession himself at the material time.[118] These comments can be seen as incorrect given that they failed to properly consider whether the effect of the notice was to deprive the seller of legal possession of the estate, and the right to factually possess the estate (the constituent elements of the obligation to give vacant possession). The learned judge failed to properly consider

---

[116] *Davis Contractors Ltd* v *Fareham Urban District Council* [1956] AC 696, per Lord Radcliffe at 729.
[117] ibid at 729.
[118] See *Korogluyan* [1975] 30 P&CR 309, per Whitford J at 317 where it was held that 'were it not for the fact that I think the defendant's case fails on special condition 9 and general condition 6, I would for my own part have come to the conclusion that in fact at the relevant time the plaintiff was not in a position to sell with vacant possession, in the sense in which I think those words ought sensibly to be construed in the context of the whole transaction'.

the legal and factual dimensions to possession, an observation reinforced by Lord Jauncey's remarks in *E Johnson & Co* on *Korogluyan,* where he stated that:

> *whether [the] views [expressed in* Korogluyan v Matheou*] were right or wrong they threw no light on the effect of the notice in the case before [the judge], since it gave the Crown no right to* immediate possession *before or by the time fixed for completion.*

Clearly, in *E Johnson & Co* Lord Jauncey showed an appreciation that the judgment in *Korogluyan* failed to address what this chapter has identified as the central issues relevant to whether the legal right to possession, and ability to factually occupy pursuant to that right, had passed to the requisitioning authority pursuant to the notice.[119] Only a detailed analysis of the constituent elements of vacant possession is able to explain why the decision reached was incorrect; this follows from an understanding of the meaning of possession in the context of the term 'vacant possession'.

Indeed, as noted previously, possession is a term with varying meanings but, in the context of vacant possession, it has been shown to refer to both legal and factual possession. The legal possession manifest in the obligation relates to the passing of an estate in land that is vested in possession (i.e. not in interest or reversion). The factual element has been shown to relate to actual occupation of the estate transferred at the point of completion. The analysis of the above cases further underlies how the obligation to give vacant possession refers to a legal right to factual possession of an estate *on completion.*

Legal and factual issues

In all the cases discussed, the debate centred around whether the rights conferred on the acquiring authority by the relevant notice included the right to possession (rendering the obligation to give vacant possession as between seller and purchaser as being 'incapable of being performed') or whether the notice did no such thing, leaving the seller/owner free to enjoy the land and pass the right to vacant possession in accordance with the contract. These decisions which, on the surface, appear inconsistent, can therefore be understood and explained from a more informed analysis of the constituent elements of what the obligation to give vacant possession actually refers to.

Where the rights of entry and inspection conferred by section 3 of the Land Acquisition Act (Barbados) 1894 fell well short of the right of possession conferred by a requisition notice under regulation 51 of the Defence (General) Regulations

---

[119] As found in the *Sheikh* decision, when both elements are not properly considered, a perverse decision is arrived at.

1939, or by notice of entry under section 11(1) of the Compulsory Purchase Act 1969 (both of which entitle the relevant authority to dispossess the owner of the land) and rather the purchasers remained free to enter onto the land and to use it for any purpose, the contract was capable of being performed and vacant possession capable of being given at the material time. This was because the seller still had the legal right to possession of the estate to pass, and the purchaser could take factual possession at the point of completion.

As such, the obstacle posed by the service of legal notices goes to the heart of the constituent elements of the obligation to give vacant possession, in terms of the effect of the service of the notice, or subsequent actions (e.g. the handing over of keys) in taking from the seller the legal right to possession and the factual ability to occupy the estate, which the buyer must be immediately entitled to on completion. The character of the contract (in terms of the vacant possession that the purchaser contracted to acquire) changes when the right to possession and the ability to actually exercise that right can no longer be passed because of a legal obstacle preventing the delivery of either of these constituent elements of vacant possession on completion (or at the relevant time).

LEGAL OBSTACLES — THE QUESTIONS TO ASK

When an analysis of the constituent elements of vacant possession, as proposed by chapter 5, is applied to cases concerning so-called 'legal obstacles' to vacant possession, apparent conflicting decisions and inconsistencies can be understood and accounted for. The relevant determinations in cases concerning compulsory purchases and requisitioning notices concern whether the notice (or subsequent acts pursuant to that notice) pass:

- the legal right to possession, and
- the ability to actually occupy pursuant to that right (the essential elements of the obligation to give vacant possession) to the requisitioning authority.

Where they do before completion, vacant possession cannot be given at completion. Understanding the two essential elements to the obligation explains the ostensibly differing decisions reached and the reasoning for the respective judgments.

In cases where a legal obstacle is argued to present a barrier to the procurement of vacant possession, the effect of the notice or requisitioning order with respect to both the right to possession and ability to actually occupy pursuant to that right at the point of completion must therefore be evaluated. This will often be a mixed question of both law and fact given the terms and nature of the notice, the applicable statute under which the notice is served, and the specific circumstances of the case in issue/ application of the relevant facts to the given context.

Indeed, the judgments analysed were also shown to be highly contextually specific. A contemporary interpretation of the obligation in the context of the specific circumstances during the war time period, and the overriding need for the Government to have possession pursuant to the Regulations, was reflected in the differing decisions that were reached with respect to the government requisitioning of property for war time purposes. Universal to all such decisions is the fact that the Government at no time was hindered in achieving its objectives by any of the decisions concerning the procurement of vacant possession. This reaffirms that beyond the core, objective elements, there is a pragmatic and fact specific element to the vacant possession obligation which enables the obligation to be interpreted in the context of all the relevant circumstances of the contract under which the obligation arises and beyond. This further reflects the practical dimension to the vacant possession obligation.[120]

This context specific element, whilst appropriate, does however make a detailed or overarching definition of the obligation harder to achieve, and is a variable which may have an unpredictable effect on any decision reached by a Court in a particular case.

An overview of potential impediments

Over time, case law has tended to take a broad view of the proposition that any impediment which prevents the purchaser from obtaining the quality of possession for which he or she had contracted, will constitute a breach of the vacant possession obligation. Case law identifies various potential obstacles to the receipt of vacant possession, which can be divided into several different categories.

The most common example of an impediment to vacant possession is when items that should have been removed by the seller or party required to give vacant possession are left at a property on completion. There has been a plethora of case law dealing with the non-procurement of vacant possession in these terms. Case law has developed a two limb test to determine whether there has been a breach of the obligation to give vacant possession, the first focusing on the intention of the party required to give vacant possession (as manifest by its conduct in purportedly vacating the premises), with the second looking more objectively (but with reference to the specific circumstances of the case) at whether the party with the right to vacant possession can, at the point of completion, occupy the property (or a substantial part of the property) without difficulty or objection.

A second common obstacle to the receipt of vacant possession is when persons remain in the property on completion. There is a wealth of case law confirming that

---

[120] *Topfell* [1979] 1 EGLR 161, per Templeman J.

the presence of an existing tenant or other occupier at the premises on completion will prevent vacant possession being given. This is commonly because the lease is still continuing (i.e. the party has contractual or statutory rights to remain in occupation of the property) or because other persons (such as trespassers) prevent the delivery of vacant possession on completion. Case law was shown (historically) to have taken an inconsistent view when the persons in occupation had no lawful claim to possession of the property (for example, squatters), but it has been established in law that unlawful occupiers will cause there to be a breach of the obligation to give vacant possession at the material time (in a similar manner to lawful occupiers). It is also important to consider the effect of a purchaser taking possession of the property before completion, on a subsisting obligation to give vacant possession on the later completion date.

The third main type of impediment to the receipt of vacant possession will be an obstacle of a legal nature. Examples include property being compulsorily purchased or requisitioned in some way after the exchange of contracts, thus ostensibly creating a legal obstacle which prevents the giving of vacant possession on completion. Here the precise nature and form of the notice must be evaluated with reference to the legal and factual dimensions of the obligation to give vacant possession in the given context.

Whilst this chapter has explained the three types of obstacle that can constitute a breach of the obligation to give vacant possession, it is also relevant to consider what else (if sufficiently substantial) may amount to a potential barrier to the receipt of vacant possession. The next chapter examines the scope and extent of the obligation in more detail, with specific reference to the nature of the property (or land), and its state and condition, and queries whether the state and condition of a given premises can also be a (potentially fourth) barrier to the procurement of vacant possession at the material time.

Chapter 7

The Scope and Extent of the Obligation

Whilst the tests to establish a breach of the obligation to give vacant possession have been shown to be highly fact specific, it is necessary to consider what else (if sufficiently substantial) may amount to a potential barrier to the receipt of vacant possession, and therefore be relevant to the application of either limb of the tests.

Traditionally, fixtures are not seen to be relevant to the vacant possession obligation but, as this chapter will demonstrate, the scope and extent of the obligation can be argued to encompass more than just chattels (which are generally understood as the most common impediment to the receipt of vacant possession). This chapter explains how the state and condition of a property can be argued to be a barrier to the

133

Vacant Possession: Law and Practice. ISBN: 978-0-08-096680-9

procurement of vacant possession, and thus represent a fourth category of impediment.[1] In this context, the uncertainty caused by other potentially interacting contractual conditions is highlighted, and analysis from earlier chapters is drawn upon to explain how a seller may seek to rely on an 'actual state and condition' clause to escape liability when the state and condition of the property or land is argued to constitute a breach of the obligation to give vacant possession.

The chapter also considers the relevance of so-called 'lesser interests' to the obligation, including for example certain *profits*, in seeking to fully explain the scope and extent of the obligation to give vacant possession.

Status of items

The most obvious difficulty in seeking to determine the scope and extent of the obligation to give vacant possession is with regard to what status items left at the property on completion may have. Disputes can arise as to whether items left behind at a property are fixtures (and therefore part of the land) or chattels (which are personal property of the tenant obliged to give vacant possession, and which must therefore be removed). Indeed, it is commonly established that if the seller's failure to give vacant possession is due to the presence on the property of *chattels*, which affect usability of the premises, then a breach of the obligation to give vacant possession will arise if the impediment substantially interferes with enjoyment of a substantial part of the premises on completion.[2] This is one of the reasons why the distinction between fixtures and chattels has traditionally been seen to be so important in the context of vacant possession.

FIXTURES AND CHATTELS

Fixtures are physical objects which accede to the realty. Any physical object classed as a fixture as a matter of law merges with the land and title to it automatically vests in the owner of the freehold, and the object itself cannot be severed from the land by anyone other than the freehold owner.[3] Further, the purchaser of a freehold is entitled to all fixtures on the land at the date of exchange of contracts.[4] Chattels are physical

---

[1] Along with tangible impediments, persons in occupation and legal impediments (as discussed in chapter 6).

[2] See *Cumberland Holdings Ltd* v *Ireland* [1946] KB 264. Also, Megarry, W and Wade, W *The Law of Real Property* (London, Sweet and Maxwell, 7th ed, 2008) p 672 state that 'removable physical impediments' are relevant to the obligation — i.e. chattels and not fixtures which are attached permanently to the land and which pass under the contract of sale.

[3] A plethora of case law exists — see *Reynolds* v *Ashby & Son* [1904] AC 466 for example.

[4] *Taylor* v *Hamer* [2002] EWCA civ 1130.

objects which retain their independent character as 'personalty' despite close association with realty. They thus do not attach to the land and do not pass with a conveyance of the land unless stipulated in the conveyance. Therefore, a seller is perfectly entitled, and indeed obliged, to remove such items before completion. This is all based on the maxim of law *quicquid plantatur solo, sols cedit*, meaning 'whatever is affixed to the soil accedes to the soil'.[5]

TESTS TO DETERMINE

The fixtures and chattels distinction turns on two distinct but connected tests. The first test concerns the physical degree of annexation to the land. The more permanently and irreversibly the object is affixed to the land, the more likely it is to be considered a fixture. A form of gravity test for a chattel has developed out of this, in that an object that merely rests on the land due to its own weight will be classed as a chattel, and one more permanently fixed will be classed as a fixture. In *Holland* v *Hodgson*,[6] spinning looms bolted to the floor were classed as fixtures, but in *Hulme* v *Bingham*[7] heavy machinery otherwise unattached was considered a chattel. In *Botham* v *TSB Bank Plc*,[8] kitchen appliances that were only connected electrically to the land (remaining in position by their own weight) were considered chattels on this test.

Gray and Gray[9] argue that the trend in case law suggests the above test is being overtaken by a second test concerning the objectively understood purpose (or object) of the annexation. The key question in respect of this test is whether the installation of the object was intended to effect a permanent improvement to the realty or was merely a temporary addition to the realty to enhance the enjoyment of the chattel.[10] Blackburn J in *Holland*[11] gave the following example:

> *Blocks of stone placed one on top of another without any mortar or cement for the purpose of forming a dry stone wall would become part of the land, though the same stones, if deposited in a builder's yard and for convenience sake stacked on the top of each other in the form of a wall, would remain chattels.[12]*

---

[5] See Burn, EH and Cartwright, J *Modern Law of Real Property* (Oxford, Oxford University Press, 17th ed, 2006) p 156.
[6] *Holland* v *Hodgson* [1872] LR 7 CP 328.
[7] *Hulme* v *Bingham* [1943] KB 152.
[8] *Botham* v *TSB Bank Plc* [1996] 73 P&CR D1, CA.
[9] Gray, K and Gray, SF *Elements of Land Law* (Oxford, Oxford University Press, 4th ed, 2006) p 39.
[10] *Elitestone Ltd* v *Morris* [1997] 1 WLR 687, per Lord Lloyd at 690.
[11] *Holland* [1872] LR 7 CP 328.
[12] ibid, per Blackburn J at 334.

As such, both the item's degree and purpose of annexation is key in the determination of the status of an item, which will proceed on a case-by-case basis, as Burn and Cartwright state:

> *[The] question of whether a chattel has been so annexed to land as to become part of it is sometimes difficult to answer. It is a question of law for the judge, but the decision on one case is no sure guide in another, for everything turns on the circumstances and mainly, though not decisively, upon two particular circumstances, the degree of annexation and the object [or purpose] of annexation.*[13]

Supposed Importance of the Distinction

An example of the supposed importance of this distinction for vacant possession arose in the case of *Hynes* v *Vaughan*.[14] In this case, one issue surrounded a chrysanthemum growing frame and sprinkler system, and whether these could be argued to be fixtures or chattels. The sellers (as defendants) had removed these from the property after the exchange of contracts (but before completion), which was unlawful if they were fixtures as they had passed with the land to the purchaser.[15] In view of the functions of the chrysanthemum growing frame and installation of the sprinkler system, it was determined that those items could not be seen as fixtures on the property so as to pass under the contract to the purchaser. As such, they were chattels which needed to be moved in accordance with the obligation to give vacant possession. The learned judge explained:

> *The question of whether or not the defendants were entitled to remove the growing frame and sprinkler system after the contract for sale had been signed depends on whether or not the apparatus could properly be described as a fixture. If it was a fixture, the [purchaser] had contracted to purchase it along with the property, and the defendants were not entitled to remove it. If it was not a fixture, the defendants were fully entitled to remove it before completion...*[16]

The judge went on to explain that the items were considered to be chattels, given their degree and purpose of annexation to the land:

> *I do not agree ... that it is even remotely arguable that the growing frame and sprinkler system were a fixture. The function of a growing frame requires the*

---

[13] Burn and Cartwright *Modern Law of Real Property* (2006) p 156.

[14] *Hynes* v *Vaughan* [1985] 50 P&CR 444.

[15] *Taylor* [2002] EWCA civ 1130.

[16] *Hynes* v *Vaughan* [1985] 50 P&CR 444, per Scot J at 453.

frame to be movable up and down the supports as the height of the growing plants requires. The function of a growing frame requires that it be dismantled from time to time in order to enable the flower bed to be cultivated and prepared for the new seedlings. The proposition that a growing frame on a flower bed can be a fixture contradicts its function. It is, in my view, an untenable proposition. As to the sprinkler, it would be possible to have a sprinkler system with underground water pipes permanently installed, but... the sprinkler system at [the property] was attached by a rubber or plastic hose to a garden tap. How anyone, lay person or lawyer, could regard that as a fixture defeats me. It plainly, in my view, was not. On this part of the case, the plaintiff's contentions [that the items are fixtures] are not, in my view, capable of being seriously argued.[17]

The judge's determination that the items were chattels was supported by witness statement evidence of a professional nurseryman, who stated that 'it would be obvious in any event, that the growing frame was removed in order during the winter months to enable the land to be prepared for the new season's plants'.[18] In other words, the growing frame was not fixed to the land so as to pass to the purchaser under the contract. As such, the sellers were correct to remove these items and, if they had not, they would (if sufficiently substantial) have constituted a breach of the obligation to give vacant possession on completion.

This is a common issue on the sale and purchase of property. Imagine that a seller contracts to convey a property to a purchaser, the contract expressly providing that vacant possession is to be given on completion. On the morning of completion the transaction completes and the purchaser is given the keys. Later in the day the purchaser inspects the property and finds that certain items have been left by the seller. The purchaser claims that the seller was in breach of his express contractual obligation to give vacant possession and that loss has been suffered as a consequence. The seller claims that the items left were fixtures (and therefore part of the land). The proper determination of the status of the items can be seen as a preliminary issue in seeking to establish whether the items had been left behind unlawfully, and therefore constitute a breach of the vacant possession obligation (if sufficiently substantial) on completion.

The status of items (and whether they have to be removed) would therefore be relevant to the scope and extent of the vacant possession obligation and as to whether the seller may be in breach. However, whilst leftover chattels are clearly a barrier to vacant possession, there is reason to question whether *only* chattels are relevant to a breach of the obligation or whether fixtures and matters pertaining to the state and

---

[17] ibid.
[18] ibid at 454.

condition (and nature) of the property or land, may also be relevant barriers to the receipt of vacant possession. This, in turn, causes one to question whether the fixtures and chattels distinction is really all that important after all in the context of vacant possession.

State and condition

Whilst items can be classified as fixtures or chattels, it is questionable whether this distinction is relevant to the obligation to give vacant possession. For example, can items which are more akin to fixtures, and constitute part of the state and condition of a given property, ever be a barrier to the procurement of vacant possession? If so, then the distinction between fixtures and chattels becomes redundant.[19] As noted previously, it is commonly established that if the seller's failure to give vacant possession is due to the presence on the property of chattels, which affect usability of the premises, then a breach of the obligation to give vacant possession will arise if the impediment substantially interferes with enjoyment of a substantial part of the premises on completion (or at the material time).[20] There is, however, no authority on the position where the vendor's inability to give vacant possession is due to the physical state of the property.[21] It is therefore not clear if an impediment to vacant possession that is not a chattel, but more 'part and parcel' of the state and condition of the property itself, can ever cause a breach of an obligation to give vacant possession. The only case that can be argued to have some relevance to this point is the decision in *Hynes*.[22]

HYNES V VAUGHAN

As noted previously, the first issue in this case surrounded the status of a chrysanthemum growing frame and sprinkler system as chattels and not fixtures. The second issue for determination related to piles of rubbish in a garden and stable and bonfire sites; the claimants (as purchasers) complained that the presence of these constituted a breach by the defendants (as sellers) of their obligation to give vacant possession of the property. There were eight areas where material of this nature was found. Seven of

---

[19] Further, if items that are more akin to fixtures could be barriers to the receipt of vacant possession, it would also be necessary to question how any such impediment may be overridden in the context of other competing contractual conditions — see the section entitled 'State and condition as a barrier to vacant possession'.

[20] See *Cumberland* [1946] KB 264.

[21] See Harpum, C 'Vacant Possession — Chamaeleon or Chimaera?' (1998) *Conveyancer and Property Lawyer* 324, 400 (CH).

[22] *Hynes* [1985] 50 P&CR 444.

these areas were outdoors and the material included such items as rotting vegetation, plastic, string, paper, soil, pieces of timber, domestic furniture and prunings, concrete blocks, broken glass, paint tins, hardcore rubble, various timbers, corrugated iron, galvanised type wire and glass bottles.[23] The claimants contended that the presence of these various items of alleged rubbish involved a breach by the defendants of their obligation to give vacant possession of the property. This was based on the Court of Appeal decision in *Cumberland Consolidated Holdings Ltd* v *Ireland*[24] which, as noted, concerned a contract for the sale of a disused warehouse. There were cellars under the ground floor of the warehouse which had been left filled with rubbish consisting mainly of bags of cement and empty drums. Damages for breach by the defendant of its obligation to give vacant possession were awarded by the Court on the basis that such items were inconsistent with the obligation to give vacant possession.

Scott J referred to the judgment of Lord Greene in *Cumberland,* where it was argued that a general condition (stating that the purchaser was deemed to buy with full notice in all respects of the actual state and condition of the property as at exchange) could *not* modify a seller's obligation to give vacant possession with respect to chattels. Lord Greene said:

> *The rubbish forms no part of the property sold and its presence upon the property sold cannot, in our opinion, be said to be covered by [the words of the general condition] 'state and condition of the property sold'. Those words refer, in our view, to the physical condition of the property sold itself, such as its state of repair, and do not extend to the case where the property sold is made in part unusable by reason of the presence upon it of chattels which obstruct the user. Such obstruction does not affect the 'state and condition of the property' but merely its usability which is a different matter altogether.[25]*

This explained clearly that chattels were not connected to the state and condition of the property and that a general condition relating to the state and condition of the property would have *no* relevance to leftover chattels, since they formed no part of the property sold. Whilst making this distinction between the state and condition of the property sold (including fixtures thereon) and chattels, Scott J went on to note, however, that such a distinction, whilst possible in cases of the interior of buildings, was not necessarily as possible with respect to matters *outside* of the premises (such as was in issue here). In discussing Lord Greene's judgment, Scott J remarked:

> *... that the [general] condition did not protect the vendor was based on his construction of the words 'state and condition of the property sold'. Those*

---

[23] ibid at 452.

[24] *Cumberland* [1946] KB 264.

[25] *Hynes* [1985] 50 P&CR 444, per Scott J at 453. Emphasis added.

words, [Lord Greene] said, referred to the physical condition of the property sold and did not cover the presence on the property of chattels. The distinction between on the one hand the property sold and on the other hand chattels on the property can be drawn with some clarity so far as the interior of buildings is concerned. But it is a distinction which becomes blurred when applied to gardens, paddocks, stable yards or other unbuilt-on-land. And the rougher and more rural in character the land, the more difficult it becomes to draw the distinction clearly.[26]

An example was given to demonstrate this:

Take the example of piles of rubbish. All properties with house or kitchen gardens of a fair size in rural areas are likely to have at least one and often more than one rubbish pile. On to such piles will be thrown refuse from the garden. Where refuse collection has in the past been infrequent or unreliable, piles of domestic rubbish may be found. Piles of this sort will often include bits of broken glass or bits of broken furniture. Piles of ashes may be found where the debris of years of swept out fires have been dumped. Bonfire sites may be found on which combustible or mainly combustible rubbish has been placed and at regular or irregular intervals burnt. Properties with stables are almost bound to have, nearby the stables, a place where stable manure has been placed. Where the building of outbuildings, whether stables, sheds or garages, has recently taken place, there is likely to be found, pushed into some convenient corner, builders' debris, such as broken bricks, tiles or planks. These piles of rubbish are likely in an old property to be of long standing. The debris of earlier years will have become part of the surrounding earth. More recent additions may still be distinguishable. But to describe the contents of piles of rubbish such as I have described as 'chattels' and as something distinct from the property sold would in most cases be quite unreal.[27]

Scott J made clear that Lord Greene's statement of principle in *Cumberland*, with respect to the fixtures and chattels distinction, was *not* intended to deal with ordinary garden or stable rubbish which could not be distinguished from the rest of the property like everyday chattels (such as table and chairs, for example) could. Items like ordinary garden or stable rubbish as referred to by Scott J were seen to be more part and parcel of the property sold, even though they may not be affixed to the property in the way that fixtures are generally understood to be attached to the property itself. As such, the actual state and condition clause contained in the

---

[26] ibid, per Scott J at 453. Emphasis added.
[27] ibid.

contract *could* potentially have relevance with respect to these 'non-chattel like' items, which thereby caused the judge to consider the effect of condition 13(3) of the contract which provided:

> *the purchaser shall be deemed to buy with full notice in all respects of the actual state and condition of the property and, save where it is to be constructed or converted by the vendor, shall take the property as it is [that is, as it was at exchange].*[28]

Scott J considered that the condition obliged the purchaser to take the property with its existing garden and stable rubbish piles and bonfire sites (which were present on the property at exchange), because such items had merged with the actual land and become consistent with the nature of the property (even though not strictly fixtures in a traditional sense). The judge therefore explained that condition 13(3) would be relevant to such items:

> *... in my judgment, condition 13(3) does provide an answer where, first, the rubbish complained of has merged with and become part of the surrounding soil and, secondly, where the nature and extent of the rubbish complained of is consistent with the nature and character of the property sold.*[29]

As such, several of the items complained of by the purchasers were held to be covered by general condition 13(3) (some others were chattels and therefore not in issue). The judge then went on to consider what the position would have been if this was *not* the case; that is, if the items were not covered by the general condition, and applied the *Cumberland* test to determine that these items were not substantial interferences with possession in any event.

This judgment does, therefore, suggest that it is possible for matters pertaining to the state and condition of the property to *themselves* constitute a barrier to the receipt of vacant possession. That is, rubbish or piles of debris connected to the state and condition of the property, and which *cannot* properly be classified as chattels, could (according to the judge) potentially cause a breach of the obligation to give vacant possession if sufficiently substantial (and not qualified by other terms in the contract).

CONFLICTING CONDITIONS

In this context, it is necessary to consider the extent to which a general condition relating to the state and condition of the property could affect an obligation to give vacant possession. In theory, a general condition relating to the state and condition of the property would have relevance to such 'non-chattel like' items as those

---

[28] ibid.

[29] ibid.

referred to in *Hynes*. Further, where the contract is subject to a general condition that the purchaser takes the premises in the state and condition that it was in at exchange, then operation of the clause would have the effect of meaning that only *new* piles of debris or related items (which come onto the property *after* exchange of contracts) could be potential obstacles to the receipt of vacant possession. As the judge said:

> *The present case is not one in which the complaint made is that after the date of contract the vendor defendants added* new rubbish *to existing piles or created new piles of rubbish ...Condition 13(3) would, I think, provide no answer to a complaint of that sort...*[30]

This is because those items would have entered onto the property after the exchange of contracts, and condition 13(3) relates to the point of exchange. The judge continued by further indicating that chattels would not be relevant to condition 13(3):

> *...Nor is the case one in which the piles of out-of-doors rubbish of which complaint is made are in any way unusual or out of character for the type of property being sold [i.e. are chattels]. If that had been the case, it may be that condition 13(3) [relating to the state and condition of the property] would not apply.*[31]

Presumably, condition 13(3) would not apply in such a case for the reason given by Lord Greene in *Cumberland*. That reason was that chattels are not part of the state and condition of the property to which condition 13(3) has application:

> *... the condition does not relate to chattels. If the rubbish forms no part of the property sold...it cannot be said to be covered by [the words of the general condition] 'state and condition of the property sold'. Those words refer, in our view, to the physical condition of the property sold itself, such as its state of repair, and do not extend to the case where the property sold is made in part unusable by reason of the presence upon it of* chattels *which obstruct the user. Such obstruction does not affect the 'state and condition of the property' but merely its usability which is a different matter altogether.*[32]

As such, an 'actual state and condition clause' could, in principle, help to avoid there being a breach of the obligation to give vacant possession with reference to 'non-chattel like' items (that were on the premises or land at the point of exchange of

---

[30] ibid. Emphasis added.

[31] ibid.

[32] ibid. Emphasis added.

contracts), and this is addressed in the section entitled 'Actual state and condition clauses' later in this chapter.

POTENTIAL IMPLICATIONS

The decision in *Hynes* is therefore an indication that the scope of the obligation to give vacant possession may not just concern chattels, as has been traditionally perceived.[33] The case suggests that potential obstacles connected to the state and condition of the property, which are not covered by a general condition (either because the obstacles have come onto the property *after* the exchange of contracts, or the contract does not contain such a general condition), *could* be a barrier to vacant possession if they could be described as impediments which substantially interfere with the buyer's right to possession; the vacant possession tests thus needing to be applied in such circumstances. Indeed, the Court made clear that in order to succeed with their defence 'the defendants must establish…that the *state of the property* as they proposed to hand it over to the [buyer] on completion was consistent with their obligation to give vacant possession',[34] suggesting that the state and condition of the property *was* relevant to vacant possession. Indeed, reinforcing this, in conclusion the judge said:

> The state of this property of which complaint is made was, in my view, reason-
> ably in keeping with the character of the property. There has been no sugges-
> tion that it was not reasonably consistent with the state of the property at the
> date of the contract. In my judgment, the plaintiff has failed in the evidence
> she placed before me to establish any arguable case that the condition of
> the property, in the state in which the defendants proposed to hand it over
> on completion, would have involved a breach by them of their obligation to
> give vacant possession.[35]

This again indicated that certain matters relevant to the state and condition of the property could, in principle, amount to a breach of the obligation to give vacant possession. This was despite the fact that such matters would be part of the state and condition or nature of the land or property itself, and therefore *not* chattels.[36]

---

[33] ibid.

[34] ibid.

[35] ibid.

[36] Harpum 'Vacant Possession - Chamaeleon or Chimaera?' (1998) *Conveyancer and Property Lawyer* 324, 400 argues that if the items were on the property at the date of contract, and did not constitute a breach of the vendor's undertaking to give vacant possession, the purchaser was bound to take the property as it stood regardless of the clause. The fact that the judge considered whether the items *could* amount to a breach of the obligation is, however, suggestive of the items potentially being relevant to the vacant possession obligation notwithstanding that the items were more akin to the state and condition of the land.

Internal or external premises

What would appear crucial to the decision in *Hynes* is the distinction that Scott J makes between the 'inside' and 'outside' of a given premises. In *Hynes*, the 'property' that was the subject of the sale and purchase contract was the dwelling-house and surrounding land. It was clear that Scott J saw discussion in *Cumberland* as having been directed at the interior of buildings, and that such comments were not similarly applicable to determinations relating to vacant possession of external premises, land and open surroundings where potential impediments would be less easily classified as fixtures or chattels, and thus need to be assessed differently. Indeed:

> *a distinction which becomes blurred when applied to gardens, paddocks, stable yards or other unbuilt on land. And the rougher and more rural in character the land, the more difficult it becomes to draw the distinction clearly.*[37]

Scott J identified certain items which could not be properly determined as chattels but which could, if still on the property at completion, be relevant to the determination (and application of the tests) as to whether vacant possession was being given. Such items (such as rotting vegetation and pieces of timber), being more akin to fixtures given their association with the general state and condition of the external property, could therefore prevent vacant possession from being given at the relevant time.

Property law parallels

In this context, a parallel can be identified between this and claims for both adverse possession and actual occupation where the nature of the land has been seen as relevant to whether adverse possession or actual occupation had been obtained. In discussing adverse possession, Lord Browne-Wilkinson quoted Slade J in the case of *Powell* v *McFarlane*:

> *The question [of] what acts constitute a sufficient degree of exclusive physical control must depend on the circumstances, in particular the nature of the land and the manner in which the land of that nature is commonly used or enjoyed. Everything must depend on the particular circumstances...*[38]

---

[37] *Hynes* [1985] 50 P&CR 444, per Scott J at 453.

[38] *Powell* v *McFarlane* [1977] 38 P&CR 452 at 470. Emphasis added. The quotation from Slade J is a paraphrase of an often cited dictum of Lord Hagan in *Lord Advocate* v *Lord Lovat* [1880] 5 App Cas 273, per Lord Hagan at 288. See also *Red House Farms (Thorndon) Ltd* v *Catchpole* [1977] 2 EGLR 125, per Cairns LJ at 126 and the more recent case of *Port of London Authority* v *Ashmore* [2009] EWHC 954 (Ch).

The nature, or state and condition of the land, is also decisive in claims of actual occupation, where the Court of Appeal in *Lloyds Bank* v *Rosset*[39] distinguished between different properties or land:

> *[t]he acts which constitute actual occupation of a dwelling house, a garage or woodland cannot all be the same.*[40]

As such, just as the physical nature and characteristics of the subject land or property will affect the prospects of, and be relevant to, claims relating to adverse possession and actual occupation, the nature of the land to which the obligation to give vacant possession is engaged would appear to be relevant to application of the tests to determine whether a breach of the obligation has occurred. Accordingly, this further supports the analysis of the preceding chapter which highlighted the context specific nature of the application of the second limb of the *Cumberland* test, and the proposition that the objective limb of the *Cumberland* test must be interpreted with reference to the particular context in issue, taking account of the nature of the land as well. These factors, it can be argued, would appear to directly affect whether the given impediment prevents the party with the right to vacant possession from being able to occupy without difficulty or objection at completion. On the same basis, the actual state and condition of the land should be relevant to the first limb of the *Cumberland* test in respect of whether, given what state and condition the land in question is left in on completion, an intention to give vacant possession is being manifested by the party required to provide vacant possession.

PRINCIPLES

From this analysis it is possible to extrapolate some further principles that will be of assistance to judges, academics and practitioners when seeking to consider whether a breach of an obligation to give vacant possession has occurred. These principles would include:

1. Is the alleged impediment inside or outside of a building?
2. What is the nature of the land?
3. To what extent is the alleged impediment consistent with, or distinguishable from, the surrounding land?
4. When did the impediment first appear on the property subject to the contract (this is relevant to whether other conditions of the contract may affect the obligation). See section entitled 'State and condition as a barrier to vacant possession' below.

---

[39] *Lloyds Bank* v *Rosset* [1989] Ch 350. Reversed by House of Lords [1991] 1 AC 107, but these observations were not material to the later decision and remain valid.
[40] *Lloyds Bank* [1989] Ch 350, per Mustill LJ at 394.

5. What effect does the impediment have on the *use* of the property?

6. To what extent is the impediment being on the property inconsistent with the transfer of 'possession'?

The answer to these questions will assist parties in forming a view as to the likelihood that the alleged impediment will constitute a breach of the obligation to give vacant possession. As these questions are clearly relevant to interpretation of the tests to determine a breach of the obligation to give vacant possession, not considering them may potentially lead to an incorrect view being reached in any particular case.

SUMMARY

In summary, the character of the alleged or potential impediment in relation to the nature of the property or land would seem an essential element in operation of both the first and second limbs of the *Cumberland* test. The nature of the land, with reference to the alleged impediment in question, will determine the extent to which usability may be affected by the impediment to vacant possession that is complained of.

Having established the proposition that the state and condition of the property may be a barrier to the receipt of vacant possession, it is appropriate to consider in more detail whether (and if so, how) such an obligation can be modified by other contractual terms. With respect to chattels and legal impediments, an obligation to give vacant possession was found to interact with general conditions such as 'subject to local authority requirement' clauses and 'no annulment, no compensation' clauses (as was discussed in chapter 3). As noted above, in the context of the state and condition of a given property, 'actual state and condition' clauses (commonly found in residential and commercial contracts for the sale and purchase of land) can be seen to potentially have an effect on items that are more akin to the nature of the premises, and which are argued to constitute a breach of the obligation to give vacant possession. This suggests that any impediment to the right of possession (whether a fixture or a chattel) should be treated in the same terms, supporting the contention that the fixtures and chattels distinction is somewhat artificial and irrelevant in respect of vacant possession, with the crucial question relating to whether the obstacle is a substantial impediment to 'possession' at the relevant time.

Actual state and condition clauses

In the context of the state and condition of a given property (or piece of land) potentially being a barrier to the receipt of vacant possession, an 'actual state and condition' clause could become a relevant consideration in respect of whether it could modify or affect the obligation to give vacant possession.

As noted earlier, an 'actual state and condition clause' is likely to provide that:

the purchaser shall be deemed to buy with full notice in all respects of the actual state and condition of the property and, save where it is to be constructed or converted by the vendor, shall take the property as it is [that is, as it was at exchange].

EXPRESS VERSUS IMPLIED OBLIGATIONS

The discussion in chapter 3 established that whilst a vacant possession obligation can appear as an express clause in the contract, it is common for conditions to fail to cater for vacant possession expressly. Ordinarily, this will mean that vacant possession will be no more than an *implied* term of the contract. In *Cook*[41] it was held that where a contract is silent as to vacant possession, and silent as to any tenancy to which the property is subject, there is impliedly a contract that vacant possession will be given on completion. When an obligation to give vacant possession has arisen impliedly, the implied obligation will be subject to specific circumstances and to the actual knowledge of the parties. For example, where one party is aware, when entering into a contract, that the interest is subject to some impediment to vacant possession, case law suggests that if the purchaser knows that the obstacle to the receipt of vacant possession is irremovable, then the implied obligation to give vacant possession will not extend so as to include that obstacle.[42] However, if at the time the contract was made, the purchaser knew of only a *removable* obstacle then the implied obligation to give vacant possession will be deemed to include such an obstacle, and if the removable obstacle is still on the premises on completion the obligation to procure vacant possession will have been breached.[43]

The position on removable and irremovable obstructions with respect to the implied obligation to give vacant possession, can be contrasted with the position where there is an express obligation to give vacant possession. Here the position is entirely different. It has been held that an express obligation to give vacant possession *will* prevail regardless of the nature of any known potential impediment to vacant possession (removable or irremovable). In *Sharneyford Supplies Ltd v Edge*,[44] the purchaser bought land from the defendant by a contract which provided that the property was sold with vacant possession on completion. The purchaser, aware that the land was occupied, had stressed from the outset that vacant possession was

---

[41] *Cook* [1942] Ch 349 at 352.

[42] *Timmins* v *Moreland Street Property Co Ltd* [1958] Ch 110.

[43] See *Norwich Union Life Insurance Society* v *Preston* [1957] 1 WLR 813.

[44] *Sharneyford Supplies Ltd* v *Edge* [1987] Ch 305.

required and had received answers to pre-contractual enquiries from the defendant that the occupants had no right to remain in possession. The occupants refused to vacate the land on completion. The express obligation to give vacant possession meant that the defendant was in breach even though the purchaser knew, at the time the contract was formed, of an irremovable obstruction to the delivery of vacant possession (the lease).[45]

STATE AND CONDITION AS A BARRIER TO VACANT POSSESSION?

The potential for the state and condition of the property (or land) to be a barrier to vacant possession is apparent in the context of physical disrepair and also potential legal obstacles. These are discussed in turn, followed by a table that summarises the various permutations. In both cases, the possible effect of an 'actual state and condition' clause must be considered.

Physical disrepair

As discussed above, a physical impediment to possession, for example, in the form of rotting vegetation or prunings (or other items similar to those referred to in *Hynes*) which cannot be classified as chattels, but rather form part of the state and condition of the premises could, in principle, be construed as a physical obstruction to the receipt of vacant possession.

Where there exists an express special condition providing for vacant possession on completion, the purchaser's knowledge of any known impediment is immaterial.[46] Further, an actual state and condition clause would not modify an *express* special condition to give vacant possession, as the express provision to give vacant possession takes precedence and amounts to an overriding guarantee.[47] Where there is an express special condition that vacant possession will be given, a seller should thus never be able to rely on a general 'actual state and condition' clause to qualify that obligation, just as purported reliance on 'subject to local authority requirement' clauses and 'no annulment, no compensation' clauses was shown to be inappropriate in the context of an express special condition to give vacant possession in chapter 3. Whether the purported impediment was connected to the state and condition of the property, or otherwise, should be irrelevant in the face of an express special condition for vacant possession.

---

[45] See also *Hissett v Reading Roofing Co Ltd* [1996] 1 WLR 1757.

[46] *Sharneyford* [1987] Ch 305.

[47] *Topfell Ltd v Galley Properties Ltd* [1979] 1 WLR 446, per Templeman J at 450. This is also the view of Charles Harpum — see Harpum, 'Vacant Possession — Chamaeleon or Chimaera?' (1998) *Conveyancer and Property Lawyer* 324, 400.

If the condition for vacant possession is only incorporated as a general condition, then it would likely have the same status as an 'actual state and condition' clause (which normally also appears as a general condition). In such a case, it is unclear as to which clause would take precedence but, given that a special condition for vacant possession was held to override competing special conditions,[48] by analogy a general condition for vacant possession would take precedence over a competing general 'actual state and condition' clause. As such, the 'actual state and condition' clause would be of no assistance to a seller as the general condition for vacant possession would prevail.[49]

If a contract is silent as to vacant possession (and therefore vacant possession is only an *implied* term of the contract) and an impediment connected to the state and condition of the property, known to be present at the exchange of contracts, substantially interferes with and prevents enjoyment of the premises on completion and is *removable*, it would prima facie constitute a breach of the implied obligation to give vacant possession. In such a case, it is possible that the seller may try to use an 'actual state and condition' clause to exclude liability in respect of that impediment as far as the obligation to give vacant possession is concerned.[50] Logically, a general condition relating to the state and condition of the property could be relied upon by a seller to escape liability for a breach of vacant possession, on the basis that such a general condition would take precedence over only an *implied* obligation to give vacant possession.[51] The seller would rely on the general condition to claim that the purchaser is bound to take the property in the condition it was in at exchange, thereby including (and taking the property subject to) the removable obstruction or impediment. As such, an 'actual state and condition' clause could modify an obligation to give vacant possession that would otherwise be breached by a removable impediment that is connected to the state and condition of the property, by requiring the purchaser to accept the property with that impediment.

However, if the impediment connected to the state and condition of the premises, known of on exchange, is an *irremovable* obstruction, then the implied

---

[48] See *Topfell* [1979] 1 WLR 446, per Templeman J at 450 and the commentary in chapter 3.

[49] If the vacant possession clause was a general condition, and the 'actual state and condition' clause was a special condition, then the latter would have to take precedence on the basis that special conditions have greater status that general conditions, which are subordinate to special conditions.

[50] This would follow the argument and reasoning in *Hynes*.

[51] Logically, the express (general) condition will prevail on the basis that it has been expressly incorporated and that any general condition having been incorporated 'must be given its full status as a contractual term and cannot just be ignored because it is one of a number of printed conditions which the parties may well not actually have read'. See Oliver LJ in *Squarey* v *Harris-Smith* [1981] 42 P&CR 118 at 128 and chapter 3.

obligation to give vacant possession would not extend so as to include that obstacle, which would fall outside of the obligation's scope. In such a case, reliance on any other conditions would not therefore be necessary. Whether the state and condition represents a removable or irremovable obstacle is likely to be determined by whether the obstacle is capable of remedy, and will always be a question of fact and degree.[52]

See below for a table summarising the various permutations discussed above.

Legal obstacles

The potential for the state and condition of the property to be a barrier to vacant possession is apparent in the context of not just physical disrepair, but also potential legal obstacles. Indeed, an 'actual state and condition' clause may be relevant in cases with facts similar to those in *Topfell Ltd* v *Galley Properties Ltd*[53] with respect to a legal impediment to the receipt of vacant possession. As noted above, in *Topfell* the seller contracted to sell a property that was partly tenanted and partly vacant. The facilities provided on the premises were inadequate for the existing occupants. After contract, but before completion, the local authority served a notice under the Housing Act 1985, limiting the number of persons who were permitted to occupy the premises until additional facilities were provided. As a result, the vendor was unable to give vacant possession of the part of the premises that was untenanted.

If the contract contained an express special condition that vacant possession would be given on completion, then the purchaser's knowledge of the physical state of the property at the time of the contract should be irrelevant.[54] Further, as noted earlier, an express special condition to give vacant possession *will* also prevail over any conflicting contractual terms, whether special or general conditions.

If the condition for vacant possession is only incorporated as a general condition, then it would likely have the same status as an 'actual state and condition' clause (which normally appears as a general condition). As discussed above, it is unclear as to which clause would take precedence but, given that

---

[52] Given the wording of an 'actual state and condition' clause, it is unlikely that a purchaser would ever be able to claim that it did not know (or was not deemed to know) of the state and condition of the premises at exchange, and therefore the 'state and condition impediment' complained of.

[53] *Topfell* [1979] 1 WLR 446, per Templeman J.

[54] *Sharneyford* [1987] Ch 305.

a special condition for vacant possession was held to override competing special conditions,[55] by analogy a general condition for vacant possession should take precedence over a competing general 'actual state and condition' clause. This would similarly cause the 'actual state and condition' clause to be of no assistance to a seller.[56]

The position is, again, less clear cut with respect to cases where the obligation to give vacant possession is only *implied.*

Where a contract is silent as to vacant possession and the contract incorporates an 'actual state and condition' clause, then this clause could prevent a buyer arguing that the state and condition of the premises is a barrier to the receipt of vacant possession. A seller would argue that a general 'actual state and condition' clause would take precedence over an implied obligation to give vacant possession, on the basis that it is not possible to imply, into a contract, a term that is inconsistent with an express term of the contract.[57] This would again assist in cases where the 'state and condition impediment' to vacant possession was known of (or deemed to be known of) on exchange and is *removable* (where known of on exchange but *irremovable*, the implied obligation to give vacant possession would not include such an obstacle in the first place).[58]

It would clearly be advisable for the contract to provide that any impediment to vacant possession occasioned by the state and condition of the property is the concern of the purchaser; but normally this situation arises because of the contract *failing* to deal with vacant possession in the first place.[59]

Whilst there is a distinct lack of case law in this area, these are examples of how a legal impediment, connected to the state and condition of a given property, can potentially constitute a barrier to the receipt of vacant possession. The examples further highlight how any such breach of the obligation to give vacant possession on this basis can, in turn, be modified (or negated) by the incorporation of an 'actual state and condition' clause.

---

[55] See *Topfell* [1979] 1 WLR 446, per Templeman J at 450 and the commentary in chapter 3.

[56] If the vacant possession clause was a general condition, and the 'actual state and condition' clause was a special condition, then the latter would have to take precedence on the basis that special conditions have greater status than general conditions, which are subordinate to special conditions.

[57] See *Rignall Developments Ltd* v *Halil* [1988] Ch 190, per Millett J at 200. See also Oliver LJ in *Squarey* [1981] 42 P&CR 118 at 128.

[58] Again, given the wording of an 'actual state and condition' clause, it is unlikely that a purchaser would ever be able to claim that it did not know (or was not deemed to know) of the state and condition of the premises at exchange, and therefore the 'state and condition impediment' complained of.

[59] See Harpum, 'Vacant Possession — Chamaeleon or Chimaera?' (1998) *Conveyancer and Property Lawyer* 324, 400.

The table below summaries the position with respect to 'actual state and condition' clauses and express and implied vacant possession obligations:

Table 7.1 Terms for Vacant Possession and 'Actual State and Condition' Clauses

	Status of Conflicting 'Actual State and Condition' Clause
Special Condition for Vacant Possession	Always subordinate to the special condition for vacant possession, whether the 'actual state and condition' clause is a general or special condition.
General Condition for Vacant Possession	An 'actual state and condition' clause, when appearing as a special condition, would take precedence, based on the established hierarchy of terms. When appearing as a general condition, it is arguable that the 'actual state and condition' clause would be subject to the vacant possession general condition, which would take precedence.[60]
Implied Condition for Vacant Possession	As an express term, an 'actual state and condition' clause should take precedence, regardless of whether it appears as a special condition or as one of many general conditions incorporated by reference and not particularly considered by the parties. It would be incorrect for a term to be implied into a contract that was inconsistent with an expressly incorporated term. Where the 'state and condition impediment', known (or deemed to be known) of at the point of exchange, is: (a) incapable of remedy (i.e. is *irremovable*) then it would not constitute a breach of the implied obligation to give vacant possession on completion, as it would be outside of the scope of such an implied obligation; (b) capable of remedy (i.e. is *removable*) then a seller will argue that it does not have to rectify this defect before completion, on the basis that, whilst the implied obligation to give vacant possession includes *removable* obstructions, the 'actual state and condition' clause requires the buyer to take the property with the 'state and condition impediment' that was known of (or, by virtue of the clause, was deemed to be known of) on exchange, and that the buyer cannot therefore claim that the 'state and condition impediment' amounts to a breach of the seller's obligation to give vacant possession on completion.

ARGUMENTS FOR EXTENDING THE SCOPE

The analysis above suggests that the scope of the obligation to give vacant possession may not be limited to chattels as has been traditionally considered by case law and

---

[60] See chapter 3 for reasoning which motivates this proposition.

property text books.[61] Traditional views of vacant possession being concerned only with personal items may therefore need to be reconsidered in light of the comments in *Hynes*.[62] The scope of the obligation may thus encompass more than originally suggested by definitions in property text books and case law and also extend to items or impediments more akin to the fabric or state and condition of the premises or land which, if sufficiently substantial, will constitute a breach of the obligation.

Further, the fact that general conditions may affect and modify an obligation to give vacant possession with respect to the state and condition of the property, in a similar manner to chattels and legal impediments (as demonstrated in chapter 3), further suggests that any impediment which affects 'possession' (whether deemed a fixture or a chattel) is likely to be relevant to the obligation. This underlies the pervasive nature of the obligation and reinforces how the fixtures and chattels distinction may therefore be somewhat artificial in the context of vacant possession. It also reminds practitioners to be alive to the issue of the state and condition of a given premises and how a condition of the contract (perhaps only incorporated by reference) may be relevant to interpretation of the vacant possession obligation in such a context.

▍ 'Lesser interests'

The above discussion has centred on tangible/physical impediments to vacant possession, but also touched on so-called 'legal obstacles' to vacant possession in respect of how the state and condition of a given property may contravene statutory restrictions on use, thus preventing the giving of vacant possession. Compulsory purchase orders and requisitioning notices, as the main types of 'legal impediments' to vacant possession, were discussed in detail in chapter 6. Crucially, in all such cases, the analysis undertaken was set in the context of fully fledged claims, and competing restrictions, to 'possession' of the property in question. It is, however, possible to acquire or be granted less extensive rights over land which do *not* amount to 'possession'. It is relevant to consider the effect of such 'lesser interests' when interpreting the scope and extent of the obligation to give vacant possession.

DEFINITION

There is no definition, as such, of so-called 'lesser interests' but such an expression is likely to refer to interests amounting to something short of exclusive possession. An

---

[61] See for example, *Cumberland* [1946] KB 264, which distinguishes chattels from the state and condition of the property. Megarry and Wade, *The Law of Real Property* (2008) p 672 state that 'removable physical impediments' are relevant to the obligation — i.e. chattels and, presumably, not items more akin to the fabric of the land and which pass under the contract of sale.

[62] *Hynes* [1985] 50 P&CR 444 at 452.

example would be an incorporeal hereditament. Incorporeal hereditaments are burdens on an estate in land in the form of 'rights which are attached to some estate, and have become part of it, so as to be enforceable by the person in possession of it',[63] but are not themselves estates in land. One type of incorporeal hereditament is a profit. A profit in the law of real property,[64] is a non-possessory interest in land, which gives the holder the right to take natural resources such as petroleum, minerals, timber, or wild game from the land of another. Indeed, because of the necessity of allowing access to the land so that resources may be gathered, every profit contains an implied easement for the owner of the profit to enter the other party's land for the purpose of collecting the resources permitted by the profit. Whatever the type of profit (whether it be rights to graze stock, plant and harvest crops, quarry stone, sand or gravel, or take timber) in practice the exercise of that right gives the owner of it a substantial degree of control over the burdened land.[65] As such, it can be questioned as to whether such rights, while amounting to less than possession but still encumbering the estate being transferred in some way, could constitute a legal obstacle to the receipt of vacant possession, if sufficiently substantial.

Imagine that a seller contracts to convey land to a purchaser. The contract provides expressly that vacant possession is to be given on completion. Between exchange and completion a third party asserts that it has a profit over the land that will prevent development of the land by the purchaser in the manner desired. While the purchaser may have contractual remedies against the seller with respect to disclosure of third party rights, can the seller transfer the land to the purchaser on completion in compliance with the seller's obligation to give vacant possession? The third party's right is clearly an interest over the land rather than a competing claim to possession, but it prevents delivery of the property free from a claim of a right over the land (i.e. the right to pass and re-pass) that is adverse to the purchaser. The purchaser may claim that the third party's right constitutes (albeit infrequently) third party occupation of the land in some way. The purchaser might argue that the adverse right was a legal impediment that prevented it from obtaining the quality of possession for which it had contracted. If a seller's obligation to procure vacant possession does not refer to transferring the estate free from all conceivable adverse legal obstacles to enjoyment, it would be necessary to evaluate what lesser interests do and do not qualify as obstacles to the receipt of vacant possession.

---

[63] Wonnacott, M *Possession of Land* (Cambridge, Cambridge University Press, 2006) p 142.

[64] From profit-à-prendre in Middle French — 'right of taking'.

[65] Wonnacott, *Possession of Land* (2006) p 141.

HORTON V KURZKE

The law at present has not dealt in detail with how an obligation to give vacant possession is affected by intervening non-possessory interests in land that, it could be argued, act as obstacles to the procurement of vacant possession. Indeed, *Horton v Kurzke*[66] would appear to be the only case that can be seen to have relevance on this point.

This case concerned the sale and purchase of land (with vacant possession) where, following exchange, the purchaser learnt of an agricultural grazing 'right' purportedly affecting the land. The purchaser asked that completion should be deferred until after the result of arbitration proceedings to decide the agricultural grazing right claim. The seller refused and, by notice under the contract, required completion of the contract within 28 days. On the purchaser's refusal, the seller claimed that she could forfeit the deposit and resell the property. The purchaser issued a writ for specific performance of the contract, with an abatement in price if the agricultural grazing right claim should be upheld; and she later issued a summons for summary judgment. The arbitrator had meanwhile decided there was no legitimate claim for the agricultural grazing right, but the purchaser did not know that until after the issue of her summons. Completion took place and the proceedings became abortive. Therefore, the case only concerned the question of costs which were awarded against the seller given its conduct throughout the matter.

What is relevant from this decision, however, are the comments made by the learned judge as to whether the agricultural grazing right (if established) would be an issue of title, or vacant possession. Whilst the purchaser claimed that the agricultural grazing right could be a barrier to the procurement of vacant possession, Goff J was clear that this was the wrong approach:

> *The plaintiff opened her case on the footing that in the circumstances the defendant was not at any material time able to give vacant possession. I doubt whether that is an entirely correct way of approaching it. I think the real question is whether the defendant was able to prove her title. As, however, there is no sufficient evidence that the alleged claimant was in actual occupation, and the inability to give vacant possession therefore—if there were such inability—was based upon the right to possession, I think whether one looks at it as a question of vacant possession or of title, one gets back to the same position and must apply the same test.*[67]

---

[66] *Horton v Kurzke* [1971] 1 WLR 769.

[67] ibid, per Goff J at 771.

The 'same test' can be seen to be a reference as to whether the impediment/defect could be remedied by completion, and therefore vacant possession/good title could be given by the seller in accordance with the contract.

LEGAL POSITION ON LESSER INTERESTS

It would seem that the difference between the legal impediments previously discussed (such as compulsory purchase orders and requisitioning notices) and the potential legal impediment here, can be explained by reference to the nature of the right or interest. Unlike compulsory purchase orders and requisitioning notices which pass the right to possession of the property to the acquiring authority (or another party), or in the case of statutory restrictions on the user, prevent possession from being legally possible, so-called 'lesser interests' do not amount to barriers to 'possession' of the property as they are only rights *over* the land, rather than competing claims to possession of the land.

Indeed, whilst the judgment in *Horton* does not specifically discuss or explain the potential overlap between vacant possession and title, the issue with the agricultural grazing right would appear to centre on what the right specifically constituted. On the facts, it would seem that the grazing rights were more akin to a profit, and did not involve exclusive possession (i.e. clearly not a freehold or leasehold interest). In the judgment, use of the word 'tenancy' (with respect to the agricultural grazing right claimed) was intended to purportedly designate a contractual arrangement, but not an estate; indeed, the seller of the land remained the party with the right to possession which was held to have been transferred to the purchaser pursuant to the contract.

This decision suggests therefore, that so-called 'lesser-interests' are not issues of vacant possession but rather issues of title, and the case has been treated as an authority for the proposition that lesser interests will only be relevant to title, and not to the delivery of vacant possession.[68] The *Horton* decision appears to establish that the risk of a purchaser buying subject to an adverse lesser interest is an issue of *title*, thus making impediments which amount to *less* than possession not issues of vacant possession. This categorises lesser interests as distinct from the vacant possession obligation (which relates to competing claims to possession itself). This would seem logical; the scope and extent of an obligation to give vacant possession, dealing with barriers to 'possession', should not encompass *rights* which, by their very nature, do not amount to possession. Thus, legal impediments, in the form of compulsory purchase orders and requisitioning notices, can be distinguished from legal impediments such as certain profits and incorporeal hereditaments; the latter being legal rights amounting to less than possession of the land to

---

[68] Megarry and Wade, *The Law of Real Property* (2008) p 672.

which they pertain, and therefore not being relevant to the vacant possession obligation.

This discussion aids a characterisation of the scope and extent of the obligation as being concerned with *all* barriers (i.e. whether fixtures, chattels or otherwise) to 'possession', but not all conceivable rights pertaining to the land which fall short of fully fledged possession. The decision in *Horton* therefore highlights the need for close analysis of the legal impediment complained of in order to determine the scope and extent of the obligation.

Summary

The obligation to give vacant possession involves an inherently factual element: the ability to take possession in a practical sense at the point of completion. Certain tangible impediments, such as chattels or persons in occupation *will* clearly be relevant to whether the obligation has been breached, as will legal impediments which amount to competing claims or barriers to *possession* itself. Conversely, interests amounting to *less* than possession will not be, given that, by their very nature, they amount to something short of 'possession' and cannot therefore be a barrier to the receipt of 'possession' in the context of vacant possession.

There is, however, some uncertainty as to whether other obstacles can be relevant in interpreting the scope and extent of the obligation. There remains no authority on whether the state and condition of a property can itself be a barrier to the procurement of vacant possession and if so, how other contractual conditions could modify an obligation in this context. This is especially apparent where there is no express vacant possession clause and, as such, a general condition (relating to the state and condition of a given property) conflicts with only an implied vacant possession obligation.

The determination as to whether certain items constitute a breach of the obligation to give vacant possession has been shown to be made with reference to the particular context including the nature of the land or property in question. A (potentially) artificial distinction between fixtures and chattels has (it can be argued, wrongly) previously been assumed by property text books and case law to explain which items can, and cannot, be barriers to vacant possession. The analysis in this chapter suggests that the obligation can be argued to relate to *anything* that constitutes an impediment to possession, including matters more akin to the fabric or state and condition of the property. As such, and as explained with reference to the decision in *Hynes*, whilst the status of items as fixtures or chattels may not always be clear, that would not matter in this context since their classification would be irrelevant when one considers whether they are impediments to the enjoyment of the right to possession on completion (or the operative date). As such, the question is not how the obstacles should be classified (or labelled), but how substantial they are in the particular context in question.

On the basis that an impediment was relevant to the scope of the obligation, and constituted a breach, the next determination for a court will be the remedy or relief that can be awarded to the successful party. As noted previously, the remedy normally awarded to an injured party for a breach of the obligation to give vacant possession will be damages, which can often be largely unsatisfactory to purchasers which, having already paid their money before finding the property is not vacant, will be unable to occupy the property as they wish to. Chapter 8 explores the remedies available in more detail. The chapter also deals with practical timing issues relating to the receipt of vacant possession at the point of completion.

Chapter 8

Completion and Remedies for Breach

Chapter Outline

As noted in chapter 5, the fact of taking actual possession is an element of the obligation to give vacant possession which arises on completion. If completion is set for a given day (which is often the case), it can, however, be questioned as to when (during that day) completion can or must be effected. This chapter starts by addressing such practical timing issues and the implications for those who are party to a given transaction.

Whilst many transactions will complete with vacant possession being given, many also will not. On the basis that an obligation to give vacant possession has arisen and is breached by the party required to give vacant possession on completion, it must then be considered where this leaves the party which had contracted for something more than is actually obtained at the relevant time. This chapter discusses the current remedies available upon a breach of the obligation to give vacant possession and explains how, at present, they can be seen to be largely unsatisfactory

159

Vacant Possession: Law and Practice. ISBN: 978-0-08-096680-9

to the injured party. Explaining the problems, and inadequacies, of the current remedies, practical suggestions are provided as to how the obligation to give vacant possession could benefit from additional express provisions in a contract. In that regard, a definition of vacant possession arising out of the analysis in this and previous chapters is proposed, along with additional drafting suggestions to improve the position on remedies for a purchaser.

The timing of completion

Whilst an obligation to give vacant possession will arise on completion, it can be questioned as to *when* completion must take place during the day stipulated as being the 'day of completion'.

GENERAL RULE

It is not general practice to stipulate that completion is required at a specific time on the day of completion in standard sale and purchase contracts.[1] As such, the task of the Court will be to seek to give effect to the true bargain between the parties based on a fair interpretation of the contract as a whole. The effect of this, in practice, is to allow a seller to satisfy his or her obligation to give vacant possession if the purchaser secures possession at *some point* during the day of completion.

In the Canadian case of *Cooper* v *Mysak*,[2] the seller was not held to be in breach of his contractual obligation to give vacant possession merely because the tenant did not vacate the property until 9.30pm. It was held that the purchaser was not justified in refusing to complete because legal completion had not taken place at a solicitors' meeting at 4.00pm that day. In *Re Lyne-Stephenson and Scott-Miller's Contract*,[3] the property was, as far as the purchaser was aware, subject to a tenancy that was due to expire on the same day as the agreed completion date. The purchaser argued that since completion must have taken place during normal business hours, and therefore before the moment (at midnight) when in law the lease expired, he had in fact acquired the 'fag-end' of the reversion, thus entitling him to money paid to the seller

---

[1] Whilst it is not the general practice to stipulate that completion is required at a specific time on the day of completion, certain contractual consequences may flow if the money is not received by a certain time. For example, completion can be deemed to take place the following working day with the technical requirement for one day's interest on the payment of completion monies becoming due to the seller — see the section entitled 'Consequences' below.

[2] *Cooper* v *Mysak* [1986] 54 OR (2d) 346. In this case there was an express contractual obligation to give vacant possession, amounting to a special condition.

[3] *Re Lyne-Stephenson and Scott-Miller's Contract* [1920] 1 Ch 472.

by the outgoing tenant in respect of dilapidations liabilities (i.e. for wants of repair by the tenant under the lease). Such a claim was rejected by the learned judge, who held that vacant possession could be given at *any* time during the day set for completion.[4]

Whilst a seller can complete at any point during the day set for completion, it is important to note that a seller will be in breach of his or her contract if, at the moment of completion, the seller is still engaged in the process of moving out. As such, whilst completion can take place at any time on the day set for completion, it must not be effected *until* the vacant possession obligation has been complied with by the seller; this is because the obligation must be performed *at the latest* by the time completion takes place. Once completion has been effected, the right to take the rest of the day to deliver vacant possession is technically lost by the seller. Obviously, parties are unlikely to take issue with a seller needing a little time longer to move out some final items, but in theory this should have been undertaken *before* completion is effected.[5]

CONSEQUENCES

Whilst it is not the general practice to stipulate that completion is required at a specific time on the day of completion, certain contractual consequences may flow if the money is not received by a certain time. For example, completion can be deemed to take place the following working day with the requirement for one day's interest on the payment of completion monies becoming due to the seller.

Condition 8 of the *Standard Commercial Property Conditions* (2nd edition) provides:

> *8.1.2 If the money due on completion is received after 2.00pm, completion is to be treated, for the purposes only of conditions 8.3 and 9.3, as taking place on the next working day as a result of the buyer's default.*[6]

However, there is an express exclusion to this rule in the event of the seller being late, effectively preventing a seller from completing *after* 2.00pm (due to being slow

---

[4] See also the Court of Appeal decision in *Chinnock* v *Hocaoglu* [2008] EWCA Civ 1175, which held that completion at 2.44pm on the tenth working day (pursuant to a notice to complete) was not too late, with the relevant condition dealing only with the financial consequences of completing late in the day and not prescribing an actual time as a 'cut-off' for when completion was to occur on that day.

[5] Peculiarly, the *National Conditions of Sale* (20th ed) condition 5(4) entitled the purchaser to require possession to be handed over to the purchaser or the purchaser's agent on *or immediately before* the time of completion (*The National Conditions of Sale* (London, The Solicitor's Law Stationery Society Ltd, 20th ed, 1981)). From 1990, the Law Society's Standard Conditions of Sale were published (superseding the previous pre-fusion conditions) and these removed the erroneous reference to 'on *or immediately before* the time of completion', which was at odds with completion being the operative time at which the obligation to give vacant possession is engaged.

[6] *Standard Commercial Property Conditions* (London, The Law Society, 2nd ed, 2003), condition 8.

in vacating) and then seeking to claim a day's interest as a consequence of completion being deemed to take place on the next business day:

> *8.1.3 Condition 8.1.2 does not apply if:*
> *(a) the sale is with vacant possession of the property or a part of it, and*
> *(b) the buyer is ready, willing and able to complete but does not pay the money due on completion until after 2.00pm because the seller has not vacated the property or that part by that time.*

Analogous provisions are contained in the *Standard Conditions of Sale* (fourth edition) under condition 6.[7]

ISSUES FOR LAWYERS

It is true that ensuring the timing of vacant possession and completion tie in is often overlooked:

> *Although the time for performance of the obligation to give vacant possession is so closely linked with the timing of completion, there appears to be a marked tendency on the part of some solicitors to ignore this consideration when arranging a time to complete. Sound conveyancing practice dictates that when completion is effected on an agency basis, a purchaser's solicitor should instruct his agent not to complete until satisfied that the vendor has moved out. Similarly, a vendor's solicitor ought not to put his client in technical breach of contract by proceeding to complete before he has vacated the property. Yet, practitioners are known to complete without reference to their client's removal arrangements. The handing over of keys to the purchaser or his solicitor on completion constitutes a symbolic giving of possession. However, it is most unusual for a purchaser to attend completion. Purchasers have more pressing concerns to occupy them at such a time. Nor is it common practice for a vendor's solicitor to handle the keys, and certainly not when a postal completion is involved. In any event the symbolic act of handing over on completion a key to the purchaser's solicitor would be an idle ceremony of no practical benefit, if the purchaser found himself denied actual possession because the vendor had not then moved out.[8]*

The issue for solicitors in seeking to coincide completion with the giving of vacant possession was addressed in the unreported case of *Chelsea Building*

---

[7] *Standard Conditions of Sale* (London, The Law Society, 4th ed, 2003) condition 6.

[8] Barnsley, DG 'Completion of a Contract for the Sale and Purchase of Land: Part 3' (1991) *Conveyancer and Property Lawyer* 185.

Society v *Barber Young & Co.*[9] Here, the contract provided for vacant possession on completion. The purchaser's solicitors were aware that a tenant was in occupation, but in response to a preliminary enquiry had been informed that she would be vacating the flat on or before completion. The solicitors were instructed to act for the mortgage company in connection with a mortgage to a purchaser. The solicitors also acted for the purchaser. The mortgage monies were to be advanced conditionally upon the purchaser obtaining vacant possession on completion. On the day of completion, the purchaser contacted the solicitors confirming that he had obtained vacant possession. Accordingly, the solicitors authorised the purchase money (held by the seller's solicitors) to be released unconditionally. Unknown to the mortgage company and the solicitors, the seller and purchaser were acting in collusion, and the purchaser disappeared without making any mortgage repayments. When possession of the property was obtained by the mortgage company, the flat was occupied by a protected tenant who had been living there for 22 years. The solicitors were sued for an apparent breach of duty in failing to ensure that the purchaser obtained vacant possession, but the claim failed. It was held that, as the solicitors had obtained the purchaser's confirmation before they proceeded to complete, they had done all that was required in the circumstances, and could not be held liable for the dishonesty of the purchaser. The judge did make clear, however, that it was always better not to finally part with the money 'until the purchaser confirms that he has obtained possession'. This seemed to the learned judge to be 'overwhelmingly the safest course to adopt, because the one person crucially concerned to ensure vacant possession [is obtained] is the purchaser'.[10]

The implications for solicitors arising from this decision are significant in the context of risk management on completion. The claim in *Chelsea Building Society* would be likely to have succeeded if the solicitors had not sought confirmation of vacant possession from the purchaser before completion was effected. A purchaser's solicitor is therefore well advised to ask their client to confirm that there is, or will be, no impediment to obtaining vacant possession on completion, especially when monies have been advanced from a lender with specific conditions associated with their release. Given how uncommon it is for a purchaser's solicitors to seek confirmation in these terms, it is rather surprising that there are not more professional negligence claims arising out of completion being effected before vacant possession has been given.

In relation to the handing over of keys on completion, the parties frequently agree that the purchaser can, on the day of completion, collect the keys from the vendor's

---

[9] *Chelsea Building Society* v *Barber Young & Co* Ch D, 7 February 1990. The facts and judgment are taken from the Lexis transcript.

[10] It is commonly the case that the purchaser's solicitor expressly stipulates that the transmitted funds are held to the purchaser's order pending completion.

estate agent, or the builder's site office (in the case of a new house), or from a friendly neighbour. A prudent solicitor, aware of such an arrangement, will instruct the seller to advise the neighbour or other agent *not* to hand over the keys until authorised to do so by the solicitor. Such authorisation should be withheld until completion has taken effect with the receipt of vacant possession having already been confirmed.

Post Completion

As noted, vacant possession must be provided to a purchaser by the time of actual completion; as such, a seller cannot be charged with a breach of its obligation to give vacant possession if the obstruction to vacant possession arises for the first time *after* completion. This point arose as a secondary issue in *Sheikh* v *O'Connor*,[11] which was discussed in chapters 5 and 6. Here, the evidence indicated that the occupant (a tenant) did not change the locks, so barring the purchaser's entry, until *after* the sale had completed (the locks were changed either during the evening of the day of completion or later). As such, vacant possession had already been given (at the point of completion earlier that day) and actions taking place *thereafter* were the responsibility of the purchaser, as new owner.

If completion takes place in the morning of the day set for completion, and in the afternoon trespassers unlawfully take possession of the property, then that will be an issue for the new owner of the property. The seller's obligation, arising at the point of completion, is discharged from then onwards. Issues can arise when it is uncertain as to when a given impediment arose, or whether an impediment arises because of acts before completion (i.e. the premises not being secured properly). All this underlies the need to ensure that the property is checked before completion to ensure that vacant possession, at that point in time, is given; completion can then be safely effected.

Remedies available for breach

On the basis that an obligation to give vacant possession has arisen and is breached by the party required to give vacant possession on completion, it must then be considered where this leaves the party who had contracted for something more than is actually obtained at the relevant time.

If, on the day of completion (but before completion is effected), a purchaser was to inspect the premises and see that they were not vacant, the purchaser would have the options outlined in the following sections.

---

[11] *Sheikh* v *O'Connor* [1987] 2 EGLR 269.

Seek Specific Performance

A purchaser would be entitled to apply to the Court for an order for specific performance, and claim damages for the impediment.

Specific performance is an equitable remedy available, at the discretion of the Court, in cases where there has been a breach of contract (i.e. of a contractual term providing for vacant possession). Effectively, specific performance can be seen to be a decree by the Court which forces or compels the party in breach to perform its contractual obligation (i.e. to give vacant possession). It is commonly sought by purchasers of land,[12] indeed:

> ... *land is always treated as being of unique value, so that the remedy of specific performance is available to the purchaser as a matter of course.*[13]

Specific performance can be granted in addition to or instead of damages. Whilst damages which are available as a matter of right, specific performance (being an equitable remedy) is only available at the discretion of the court, such discretion exercisable by settled principles of law.[14]

It is also possible for a purchaser to seek specific performance with an *abatement* of the purchase price to reflect the deficiency in question.[15] In *Basma* v *Weekes*,[16] a contract to sell land from three tenants in common was defective given that one of the co-sellers lacked capacity. The purchaser was successful in obtaining two shares, which were capable of being conveyed, with compensation to reflect the share that could not be conveyed.[17]

In cases of vacant possession, either of these principles should apply with a purchaser seeking specific performance of the contract less an amount to reflect the impediment to vacant possession (most likely if such a defect is irremovable) or damages for having to remedy the impediment (most likely in cases where the defect is removable) (see below for more on damages).

In terms of the duties of a seller subject to an order for specific performance, according to *Wroth* v *Tyler*[18] a seller will not normally be obliged by an order for specific performance to undertake 'hazardous' litigation to obtain possession, but

---

[12] See Megarry, R and Wade, W *The Law of Real Property* (London, Sweet and Maxwell, 7th ed, 2008) p 690.

[13] See *AMEC Properties Ltd* v *Planning Research Systems Plc* [1992] 1 EGLR 70 at 72.

[14] ibid.

[15] See *Rutherford* v *Acton-Adams* [1915] AC 866 at 870.

[16] *Basma* v *Weekes* [1950] AC 441.

[17] Further, under s 50 of the Supreme Court Act 1981, a purchaser may instead seek specific performance against a seller with damages in addition. Megarry and Wade, *The Law of Real Property* (2008) p 690 report that 'the precise interrelationship between these alternative remedies is obscure'.

[18] *Wroth* v *Tyler* [1964] Ch 30.

would still remain liable in damages. It was held in this case that a seller who sold with vacant possession had, if necessary, to take proceedings against any wrongful occupant but he would not usually be required to embark on 'difficult or uncertain litigation'.

SERVE A NOTICE TO COMPLETE

Rather than seeking to enforce the terms of a given contract, a purchaser may prefer to refuse to complete and recover any deposit paid, in addition to suing for damages for loss of bargain. To do this a purchaser will normally need to serve a 'notice to complete' on the seller and after expiry of that notice (which will be determined by contractual provisions) then rescind the contract, recover any deposit paid and claim damages.

Contracts can include an express provision that 'time is of the essence of an agreement'. The effect of this would be to allow the party relying upon the clause to immediately terminate the agreement and, if appropriate, claim damages if the other party fails to perform an obligation in accordance with the date or time specified in the agreement. However, time is not normally of the essence in sale and purchase contracts. As such, whilst a breach of the obligation to give vacant possession may occur on completion, a purchaser does not have an automatic right to terminate the contract there and then (which can be contrasted with a failure to give good title or disclose latent defects, for example). As such, the notice to complete must be served requiring the contract to be completed, normally in accordance with condition 8 of the *Standard Commercial Property Conditions* (second edition) or condition 6 of the *Standard Conditions of Sale* (fourth edition), and in order to make 'time of the essence', thus allowing rescission in the event that it is not completed within the specified timeframe.

Condition 8 of the *Standard Commercial Property Conditions* provides:

8.8   Notice to complete
8.8.1   At any time on or after completion date, a party who is ready, able and willing to complete may give the other a notice to complete.
8.8.2   The parties are to complete the contract within ten working days of giving a notice to complete, excluding the day on which the notice is given. For this purpose, time is of the essence of the contract.[19]

These conditions are replicated under condition 6 of the *Standard Conditions of Sale* (fourth edition), with an additional third clause relating to making up any shortfall in the 10% deposit:

6.8.3   On receipt of a notice to complete:
(a)   if the buyer paid no deposit, he is forthwith to pay a deposit of 10%;

---

[19] *Standard Commercial Property Conditions* (2nd ed, 2003) condition 8.8.

*(b)   if the buyer paid a deposit of less than 10%, he is forthwith to pay
a further deposit equal to the balance of that 10%.[20]*

The notice to complete allows the party who has served the notice to rescind the
contract upon expiry of the notice, in the event that completion has not taken place.

Where the notice is served by a buyer and not complied with by a seller, upon
rescission of the contract the buyer may, in addition to claiming damages for breach
of contract, recover the deposit plus interest, as provided by condition 7.2 of the
Standard Conditions of Sale (fourth edition):

7.2 Rescission
If either party rescinds the contact:
*(a) unless the rescission is a result of the buyer's breach of contract the
deposit is to be repaid to the buyer with accrued interest;*
*(b) the buyer is to return any documents he received from the seller and is to
cancel any registration of the contract.[21]*

This is replicated by condition 9.2 of the *Standard Commercial Property* Conditions.

It is unlikely that a seller would serve a notice to complete for an issue relating to
vacant possession, as it is for the seller to give vacant possession and, if vacant
possession is not being given, the seller would not be 'ready willing and able' to
complete, thus making it impossible for it to serve a valid notice on such a basis.
With that said, it is technically possible for a seller to serve such a notice if, for
example, it considered that the premises *were* ready to be transferred with vacant
possession but that was disputed by the buyer, who was refusing to complete. In such
a case, the buyer would run the risk of the contract being rescinded by the seller
(upon expiry of the notice) and the deposit retained by the seller.

COMPLETE AND CLAIM DAMAGES

Even if vacant possession was not given on completion, a purchaser can choose to
complete without prejudice to a right to claim damages.

The availability and quantum of damages will be determined by the circumstances and the nature of the losses in question. A purchaser's remedies may also be
restricted by the express terms of the contract.[22] However, as a general rule, a
purchaser is likely to be able to recover as damages the sum necessary to place him or
her in the position in which he or she would have been had the contract been
performed.

---

[20] *Standard Conditions of Sale* (4th ed, 2003) condition 6.8.3.

[21] ibid, condition 7.2

[22] See PLC Property, 'Selling with Vacant Possession' (accessible via subscriber service).

Where the impediment to vacant possession is irremovable, for example if the property is let to a protected tenant, the measure of damages will be the difference between the purchase price and the market price of the property subject to the impediment. It will also be possible to claim for consequential losses. For example, in *Beard* v *Porter*,[23] costs arising from the purchase of another property were held to be recoverable.

Where the impediment is removable, then the purchaser may recover the cost of actually removing the impediment to vacant possession. In *Cumberland Consolidated Holdings Ltd* v *Ireland*,[24] the purchaser was successful in recovering the cost of removing rubbish left in the premises. Whether an impediment is removable or irremovable will be a question of fact and degree.

Problems in practice: already completed?

The remedies outlined above are all open to a purchaser in the event that an inspection of the property takes place *before* completion. Commonly, however, a given property is not inspected prior to completion, even though this (as discussed above) is best practice. It is stated on *Lexis Nexis Butterworths Direct* that ideally:

> the buyer's conveyancers should check for any evidence as to rights of occupiers by either personally inspecting the property or advising the buyer client to do so...the buyer's conveyancers should raise a requisition of the seller's conveyancers requesting confirmation that vacant possession of the whole of the premises will be given on completion and that all occupiers have agreed to vacate.[25]

In practice, this does not normally occur. The first a purchaser knows about the problem with vacant possession is after completion when the purchaser arrives at the premises to find that all is not as expected. At this point, the contract has been completed (the seller has the sale monies in cleared funds) and the purchaser is left having to claim damages for a property that it cannot immediately occupy as it wished to. This is largely unsatisfactory. Further, it leaves the purchaser with the burden of having to advance a claim to recover the loss sustained as a consequence of the breach of the vacant possession obligation which may prove difficult, or impossible, if (for example) the seller has weak 'covenant strength' (i.e. is not worth suing). If the obstacle to vacant possession is a person with a right to remain in

---

[23] *Beard* v *Porter* [1948] 1 KB 321.

[24] *Cumberland Consolidated Holdings Ltd* v *Ireland* [1946] KB 264 at 270.

[25] [547] 10 Occupiers (accessible via subscriber service).

occupation, the purchaser may have difficulty in removing them from the premises and therefore may have to take the premises subject to their interest.[26]

NON MERGER

There is authority for the proposition that the breach of an obligation to give vacant possession gives a purchaser the right to choose to rescind the contract even *after* completion. This is because the obligation to give vacant possession has been said not to merge in the conveyance or transfer, but to remain actionable after completion (even in the absence of an express non-merger clause). This, however, is subject to the purchaser having not affirmed the contract.

In *Hissett* v *Reading Roofing Co Ltd*, [27] the defendants agreed to sell to the first plaintiff property comprising offices, depot space and a flat; the property was sold subject to a special condition that vacant possession be given on completion. Condition 33 of the Law Society's Conditions of Sale 1953 stated that:

> *notwithstanding the completion of the purchase any General or Special condition or any part or parts thereof to which effect is not given by the conveyance and which is capable of taking effect after completion ... shall remain in full force and effect.*[28]

The contract was completed and on the direction of the first plaintiff it was transferred to the second plaintiffs, a company. The plaintiffs were unable to get vacant possession of the whole property because the flat was at all material times occupied by a protected tenant. The plaintiffs claimed damages for breach of the sale agreement. It was held that the first plaintiff was entitled to damages for breach of contract because the defendants failed to give vacant possession in accordance with the special condition which was (in the words of condition 33) a condition 'to which effect' was 'not given by the conveyance' and was 'capable of taking effect after completion'; further the condition did not merge with the conveyance which incorporated only part of the terms contained in the contract for sale.

In *Gunatunga* v *Dealwis*,[29] it was noted that there was established authority for the proposition that a contractual term that vacant possession shall be given on

---

[26] For a discussion of the doctrine of constructive notice (with respect to overriding interests and other adverse interests to which a sale may be subject), see Howell, P 'Notice: A Broad View and a Narrow View' (1996) *Conv* 34; Partington, D 'Implied Covenants for Title in Registered Freehold Land' (1988) *Conv* 18 and Sheridan, D 'Notice and Registration' (1950) *NILQ* 33.

[27] *Hissett* v *Reading Roofing Co Ltd* [1969] 1 WLR 1757. In *Hissett* the obligation was express but the result should be the same even if the term for vacant possession was implied.

[28] *The Law Society's General Conditions of Sale 1953* (London, The Law Society, 1953) condition 33.

[29] *Gunatunga* v *Dealwis* [1996] 72 P&CR 161.

completion did not merge in the conveyance. In that case the respondent's conduct post-completion, seeking to run the business in order to prevent its collapse and the loss of its goodwill, was *not* held to amount to affirmation of the contract. The failure by the appellants to give vacant possession on the relevant date gave rise to a new and separate cause of action to rescind the contract post-completion.

As noted above, in practice, by the time a purchaser becomes aware of the breach of the vacant possession obligation, a period of time (sometimes a number of days) will have passed and the purchaser may have commenced using the premises and can therefore, by conduct, be deemed to have affirmed the contract (although as noted in *Guntunga* this will normally be a question of fact and degree given the circumstances of the case). Even then, with the monies having been transferred over to the seller to effect completion, this leaves the purchaser having to embark on legal proceedings to seek to action the breach and seek to have the purchase monies returned. Whilst legally actionable in this way, the non-merger provision is unlikely to be of use to a party seeking to rescind the contract when they find that vacant possession has not been provided on completion having already paid over the completion monies. Normally, a purchaser would advance a damages claim instead (see the section entitled 'Complete and claim damages' above).

Better contractual wording

At present, the current law and practicalities of completion (which more often than not take effect, by convention, *prior* to an inspection of the property) put the seller in a much stronger position, as far as a breach of a vacant possession obligation is concerned. A purchaser will often be left in the difficult position of advancing a claim for damages, having suffered interruption as a consequence of not being able to immediately occupy without difficulty or objection, or may have lost a proposed subletting opportunity. The purchaser can sometimes suffer even greater detriment if it had already contracted to demise the premises to a tenant on the basis that a transfer with vacant possession would take place. This can result in the purchaser being subject to breach of contract claims (from the proposed tenant), giving rise to consequential losses.

As noted earlier, and discussed in more detail in the next chapter, in the leasehold context, the landlord currently has the upper hand and can use the issue of vacant possession to seek to prevent the tenant exercising a contractual break, option in a lease if the landlord would prefer the lease to continue. This is not just when vacant possession is an express condition of lawful operation of the break, but also in circumstances where the break is conditional upon material compliance with covenants which, by virtue of the yielding-up obligation, will include a requirement to give vacant possession in any event.

It is therefore appropriate to suggest ways to potentially improve this state of affairs. Development of the concept of vacant possession should not, however, be limited to the remedies that are available when a breach has occurred, but also more generally with reference to the term itself. Indeed, a means by which the uncertainty and misunderstandings caused by the term, given its inconsistent evolution and understanding in case law, can potentially be redressed, is to provide a better definition of the obligation as a defined contractual term itself.

DEFINITION OF VACANT POSSESSION

Currently, both the *Standard Conditions of Sale* (fourth edition) and the *Standard Commercial Property Conditions* (second edition)[30] fail to provide a definition of vacant possession. As discussed in chapter 4, they both deal with vacant possession as a special condition, but without explaining what is meant by the term:

The property is sold with vacant possession on completion
OR
The property is sold subject to the following leases or tenancies...[31]

The property is sold with vacant possession on completion
OR
The property is sold subject to the leases or tenancies set out on the attached list but otherwise with vacant possession on completion.[32]

From the comprehensive review of the obligation undertaken in the preceding chapters, it is suggested that use of the term 'vacant possession' in standard sale and purchase contracts, which incorporate either set of the current Standard Conditions of Sale, could be appreciably enhanced by providing vacant possession with a defined meaning. This would assist in allowing the obligation to be interpreted more satisfactorily in the given context in which it becomes a salient issue, or point of contention.

A definition of vacant possession, which could therefore be used in standard sale and purchase contracts and in the current editions of the Standard Conditions of Sale, with a capitalised 'V' and 'P', would be as follows:

'Vacant Possession' means:
The Buyer [or party with the right to vacant possession] being able to actually enjoy their right of possession immediately on Completion without any form of Substantial Impediment to that right of possession.

---

[30] *Standard Conditions of Sale* (4th ed, 2003) and *Standard Commercial Property Conditions* (2nd ed, 2003).

[31] *Standard Conditions of Sale* (4th ed, 2003) special condition 4.

[32] *Standard Commercial Property Conditions* (2nd ed, 2003) special condition 3.

In this definition, 'Impediment' means:

An object or issue (whether physical or legal) which prevents or interferes with the Buyer [or party with the right to vacant possession] being able to occupy the whole or a Substantial part of the Property without Difficulty or Objection.

In this definition, what constitutes 'Substantial', 'Difficulty' and 'Objection' are each questions of fact and degree, to be objectively assessed in the specific circumstances of the given case at completion [or the material time] and taking into account (a) the nature/state and condition of the premises, property or land subject to the obligation; (b) the characteristics of the parties in question; and (c) the current or (if different) intended use of the property (at completion).

This reflects judicial comment in case law, an analysis of the legal and factual dimensions to the obligation, and a practical awareness of the circumstances in which the obligation is likely to manifest itself in commercial contexts. As such, the definition reflects the fact that, whilst certain parts of the vacant possession definition can clearly be articulated, other elements will be fact specific determinations in each given case. This clearly improves on current understanding of the obligation by actually prescribing the relevant variables that should be taken into account when interpreting the tests in a specific context. Providing a definition of the term as part of the *Standard Conditions of Sale* would provide greater certainty for the parties and a better understanding of how the tests to determine a breach of the obligation will operate in real life scenarios. Practitioners are well advised to include this (or a related) definition in the 'special conditions' section of sale and purchase contracts in order to give the term defined meaning and understanding in the context of the contract more generally.

A more clearly defined contractual definition would also assist in the leasehold context, in respect of the requirement to give vacant possession at the termination of a lease (whether that be by effluxion of time, or pursuant to the exercise of a break right). This is discussed in further detail in the next chapter.

IMPROVED REMEDIES

In addition to providing a clearly articulated definition of vacant possession, a contract for the sale and purchase of land would also benefit from amendment in respect of the remedies that may be available upon a breach of the obligation. This could, in turn, be used to rebalance the risk and responsibility between the parties which, as discussed above, is presently in favour of the seller at this time, given the practicalities and conventions associated with completion.

The current remedies associated with the obligation could be improved or ameliorated in the following ways.

Make time of the essence

In contracts where vacant possession on a particular day is *absolutely* fundamental to the contract (as opposed to where it is just important), it may be appropriate to seek to make time of the essence in respect of the vacant possession provision. The effect of this will be to provide the parties with immediate remedies on the day of completion if vacant possession is not given (i.e. rescission without the need to first make time of the essence by serving a notice to complete).

A seller may seek to reject such a provision, because making time of the essence for a vacant possession obligation will clearly impose on the seller greater risk, and thus a burden, of ensuring that there is no conceivable argument that vacant possession has not been given, in view of the immediate rights of the buyer from the day set for completion in the event that vacant possession is not given on that day.[33] It would, however, have the effect of focusing the seller's attention on the procurement of vacant possession on completion, and provide the buyer with greater comfort that they are likely to receive what they have contracted for on completion (or the immediate right to discharge itself thereafter if not). If the seller is promising that 'vacant possession' will be given, then it can be argued that they should be prepared to deal with potentially harsher and more immediate consequences in the event that it is not.

Reduced time for notice to complete

Under the *Standard Conditions of Sale* and *Standard Commercial Property Conditions*, a notice to complete must allow 10 working days in order to give the party the right to thereafter rescind the contract.[34] If 10 working days are not provided for by the notice, then the rescission could itself be deemed unlawful, giving rise to legal proceedings.

There is, however, nothing to stop the parties amending the relevant general condition and providing that only five working days need to be given on the service of a notice to complete (whether generally or specifically with regard to the issue of vacant possession). This can be useful in cases where a proposal to make time of the essence in respect of the vacant possession obligation (as suggested above) is refused by a seller. Requiring a notice to complete to only prescribe five days expedites the time at which the parties' remedies and actions can then take effect, thus still resulting in the seller needing to give greater attention to giving vacant possession.

---

[33] Obviously, the seller has until midnight on the day of completion to provide vacant possession unless a specific time that day, by which vacant possession must be given, is expressly stipulated.

[34] See *Standard Conditions of Sale* (4th ed, 2003) condition 6.8 and *Standard Commercial Property Conditions* (2nd ed, 2003) condition 8.8.

The effect of this is to rebalance the risk and responsibility of the parties a little more (but not completely) in favour of the buyer.

Liquidated damages

Whilst case law on damages explains, in principle, what can be claimed by an injured party in the event that there is a breach of the obligation to give vacant possession, if parties are aware (from the outset) of the potential losses that could be incurred by a given party, then it may be worth considering expressly providing for certain pre-agreed damages if vacant possession is not given on completion.

Liquidated damages are a fixed or pre-determined amount which the parties can agree will contractually become payable upon a given breach of the contract. The issue with purported liquidated damages clauses, however, is that the amount stipulated must be a genuine pre-estimate of loss to be sustained in a particular circumstance. If not, the clause is likely to be seen to constitute a penalty, thus rendering it unenforceable.

If, for example, a buyer requires vacant possession of a property in order to complete on an agreement for lease which it has entered into with a prospective tenant, which itself is conditional on the buyer (as proposed landlord under the agreement for lease) giving vacant possession of the said property, then a liquidated damages clause could be useful. In this and similar such circumstances, it would be possible to estimate the likely loss of losing the proposed tenant (on a weekly basis) and thereafter any difference in value between securing the letting that was the subject of the agreement for lease, and securing an alternative tenancy at perhaps a lower rent. In similar terms, if vacant possession of a property is required for a specific purpose or event, then the costs of having to relocate the event or change the arrangements are likely to be capable of quantification in advance, thus making a liquidated damages clause possible.

Obviously, if a seller contracts to pay specified damages in the event of not procuring vacant possession on completion, the seller is more likely to take active steps to ensure compliance with all obligations in this regard. It will, of course, depend on the wider context of the overall transaction as to whether a seller would be prepared to allow a liquidated damages clause to enter into the contract. In such cases, a seller may be well advised to consider insuring against the additional risk that arises from such a potentially onerous contractual provision becoming operative due to a failure to provide vacant possession.

Deemed inspection pre-completion

Putting aside the arguments of non-merger of a term for vacant possession in a contract, a purchaser's position is appreciably undermined if completion takes place *before* an inspection of the property has revealed whether the premises are vacant. One view is that, whilst the seller is the party responsible for ensuring that vacant possession is given, a buyer should take on *some* of the responsibility for ensuring that what it has contracted for is conveyed to it on completion and before the monies are paid over.

Parties to a contract could first provide that an inspection (revealing no impediment to vacant possession) is 'deemed' to have taken place before completion and second that no claim for a breach of the obligation to give vacant possession can be made thereafter (in respect of impediments that would have been reasonably discoverable from such an inspection). It is likely that tangible impediments or persons in occupation would be those that a clause of this kind would most probably relate to, as legal impediments (e.g. a Housing Act notice) are unlikely to be discoverable on a physical inspection of the property.

This has particular advantages to the seller because it ensures that, once completed, the risk of non-procurement of vacant possession has passed to the buyer, with no arguments or claims thereafter being capable of being advanced by the buyer (in respect of matters that were reasonably discoverable on such an inspection). Such a clause would obviously encourage a buyer to undertake a reasonable inspection to ensure that vacant possession has been given (as far as can be determined) before completion takes effect, something which should happen but often does not. In practical terms, however, especially in residential transactions, inspections at the point of completion can be impossible or impractical; as such, inclusion of such a clause is more likely to tilt the balance even more in favour of sellers who already can be seen to have the upper hand as far as vacant possession is concerned. Indeed, deeming an inspection to have taken place before completion (revealing no reasonably discoverable impediments) can be seen to amount to a contractual means of partly negating the covenant for vacant possession itself, as the buyer is thereafter partly waiving its rights to argue otherwise. With that said, it would provide contractual certainty to a seller that completion really does mean an end to its vacant possession obligations under the contract, so far as reasonably discoverable impediments are concerned.

Summary

The fact of taking actual possession is an element of the obligation to give vacant possession which arises on completion. If completion is set for a given day (which is often the case), it can be questioned as to when during that day completion can or must be effected. This chapter has explained how, because it is not general practice to stipulate that completion is required at a specific time in standard sale and purchase contracts, the effect of this is to allow a seller to satisfy its obligation to give vacant possession if the purchaser secures possession at *some point* during the day of completion. The practical and professional problems that this can create have been identified, with strategies to reduce and eliminate risk as far as possible.

On the basis that an obligation to give vacant possession has arisen and is breached by the party required to give vacant possession, it must be considered where this leaves the party who had contracted for something more than is actually obtained at the relevant time. This chapter has explained the rights and remedies of both seller and buyer in the event that vacant possession is not given at completion. In doing so, it has also provided practical suggestions as to how to improve the position on remedies for parties to a contract, and also proposed a clearly defined and articulated definition of the contractual obligation to give vacant possession itself. All these seek to add further clarity to practical operation of the obligation in real life scenarios.

Whilst this and previous chapters have focused on the obligation to give vacant possession in the freehold context, the principles are equally applicable to the leasehold context. Chapter 9 specifically focuses on issues of vacant possession in the leasehold context, with practical points to note for lawyers and other professionals.

Chapter 9

Vacant Possession in the Leasehold Context

Chapter Outline

Earlier chapters focused on the problems associated with vacant possession in a freehold context and, in particular, with reference to sale and purchase contracts. The most common situation in which vacant possession is relevant in the leasehold context is with respect to the grant and termination of business leases over properties, **177**

Vacant Possession: Law and Practice. ISBN: 978-0-08-096680-9

and difficulties caused by not giving vacant possession can be even more onerous here. For example, a tenant may wish to bring its lease to an end before the contractual expiry by exercise of a break option. The exercise of the break can be contingent upon the procurement of vacant possession, a condition which, if not met, can cause the break to be ineffective and the lease to continue.

This chapter explains how an obligation to give vacant possession can arise, and be breached, in the leasehold context, especially with reference to conditional breaks, and the issues which need to be addressed in such circumstances. It also looks at other leasehold issues to which vacant possession will be relevant, including the assumption of vacant possession on rent review.

When will the obligation arise?

Vacant possession is an essential element of any land transaction where the right to occupy a property is being vested in, or passed to, a third party.[1] In the leasehold context, vacant possession will be relevant to the grant, transfer and termination of leases and other tenancies (and perhaps informal agreements to occupy by consent).[2]

GRANT

If the lease is to be 'vested in possession' it will be implied (as part of the grant) that vacant possession is being given. If the lease was to be vested only 'in interest' (i.e. a reversionary lease that is granted to begin at some time in the future, usually after the prior existing lease has expired), then vacant possession would not be implied as it would not be intended to give an immediate right to enjoy the estate in land.[3] The right to enjoy the estate, in such a case, would be postponed to some future date, when its term would start.

If a tenant is granted a lease only to find that it cannot take up occupation, then a tenant would be likely to advance a claim against the landlord for derogation from

---

[1] The only category of transaction for which vacant possession will not be relevant concerns the transfer of reversionary interests (for example, a freehold subject to a long lease) and other estates that are not 'vested in possession'.

[2] This is determined by what definition of vacant possession is adopted, as discussed in chapter 1. Upon termination of a licence a tenant will normally be obliged to vacate the premises, but it is arguable as to whether this is giving vacant possession as the tenant is not returning possession (like, for example, on termination of a lease). The essence of a licence, as opposed to a lease, is that it does not involve the grant of possession and this is what distinguishes it from a lease. Accordingly, as such, it does not amount to a legal estate or interest in land and does not bind third parties, but it is common for licences to make use of the term.

[3] See *Long* v *Tower Hamlets LBC* [1996] 2 All ER 683.

grant and/or for breach of the landlord's express or implied covenant for quiet enjoyment.

If the lease is pursuant to an agreement for lease (perhaps conditional on such matters as planning permission being granted or landlord's works), then the agreement for lease is likely to incorporate (by reference) the *Standard Commercial Property Conditions* (second edition); unless amended, these will include the giving of vacant possession upon the grant of the leasehold interest as a special condition.[4]

TRANSFER

In similar terms, on the assignment of a leasehold interest it will be implied that the assignee will be able to take possession of the property pursuant to the assignment (unless the contrary is intended). If the assignment is pursuant to an agreement to assign, this may also incorporate (by reference) the *Standard Commercial Property Conditions* (second edition) which will similarly include vacant possession as a special condition, unless an amendment to the conditions is made. It is because there is not usually a contract on the assignment of a lease that vacant possession will often not be dealt with expressly, unlike in the context of the sale and purchase of freehold land.

TERMINATION

If a lease runs for its entire term, then giving vacant possession will become relevant to the return of possession to the landlord when a lease comes to an end by effluxion of time (and no statutory or common law rights to remain in the property are engaged).[5] The obligation will also arise on an earlier termination due to, for example, the exercise of a break right.

In relation to the exercise of break rights, the obligation may arise following a lease ending pursuant to the tenant (or landlord) having given the requisite notice. In other cases, however, giving vacant possession may itself be an express precondition for exercise, by the tenant, of the break right (that is, vacant possession will be part of the conditionality associated with the break actually taking effect), or may be an implied requirement by virtue of other covenants in the lease, as the next section explains.

---

[4] See *Standard Commercial Property Conditions* (London, The Law Society, 2nd ed, 2003) special condition 3.

[5] See Bowes, C and Shaw, K 'Time's up…but I'm staying!' (2008) 218 *Property Law Journal* 9–11 and Bowes, C and Shaw, K 'Term of years…uncertain' (2009) 225 *Property Law Journal* 7–8 in respect of possession on the termination of leases. It is common, in the business context, for a tenant to have security of tenure under the provisions of the Landlord and Tenant Act 1954.

Vacant possession on conditional breaks

Upon the grant of a lease for a fixed term, it is common for parties to include a termination right which will allow the tenant, or both parties, to terminate the lease earlier than the contractual expiry as provided for by the lease. In law, a break clause is a right that is annexed to the lease and the reversion, and exercisable by the tenant (and/or the landlord) for the time being, and its successors in title.[6]

A break clause can probably best be understood as an 'option', which either party may 'choose' to take up, but will not be compelled or required to do so. When interpreting a break clause, its meaning will be strictly construed, and compliance with any associated conditionality will need to be strictly performed.[7] In particular, it is important to note that it will be implied that time will be of the essence in respect of time limits prescribed by a break clause.[8] Some break clauses will state this expressly.

Break rights in leases will be drafted in terms such as:

Option to Terminate
The Tenant may terminate this lease by giving not less than 6 months' prior written notice to the landlord.

The right to terminate on notice may be a right exercisable at any time (a so-called 'rolling break'), or may be with reference to, for example, an anniversary of the term of the lease (e.g. 3rd, 6th, 9th and 12th anniversaries of a 15-year term), or a specific date or dates during the term of the lease.

Commonly, a tenant's right to break a lease will not be an absolute right by the giving of notice alone (as in the above example), but rather will be conditional on one or more requirements being met by the tenant. These are known as express pre-conditions.

EXPRESS PRE-CONDITIONS

A tenant's right to break a lease may be conditional on such matters as the payment of rent or compliance with covenants under the lease. Such pre-conditions for operating a break can either be absolute or qualified.

---

[6] See *Max Factor* v *Wesleyan Assurance Society* [1995] 2 EGLR 38 and *Olympia and York Canary Wharf* v *Oil Property Investments* [1994] 2 EGLR 48. This will still be the case even if the lease does not state that references to the tenant or landlord include reference to their successors in title — see *Re 120 Packington Street* [1966] 110 Sol Jo 672. Such a rule may be overridden by express wording in the lease which provides that the right is personal to the original tenant (or landlord), or only exercisable whilst the interest is vested in them.

[7] For example, see *Friar* v *Grey* [1850] 5 Ex 584, *Bass Holdings Ltd* v *Morton Music Ltd* [1988] 1 Ch 493, *Trane* v *Provident Mutual Life* [1995] 1 EGLR 33 and *Reed Personnel Services Plc* v *American Express Ltd* [1997] 1 EGLR 229.

[8] *United Scientific* v *Burnley Corporation* [1978] AC 904.

An absolute condition, that the tenant has paid the rent and performed and observed its covenants and conditions, for example, will prevent the tenant exercising the break if there is a subsisting breach of covenant or condition at the break date, no matter how trivial the breach.[9] For example, £1 remaining due is as important as £1,000 being due, and both would have the same effect of causing the tenant to have failed to comply with a condition that all rent had been paid by the break date. In practice, vacant possession will often be stipulated as an express precondition for the exercise of a break, as an absolute covenant.

With that said, a condition may be qualified so that the tenant is required to have 'reasonably', 'materially' or 'substantially' complied with its obligations. This is most common with respect to repair obligations, for example, or the other performance of covenants. In the same terms, a vacant possession obligation could be modified and require the tenant to give 'vacant possession as far as practicable', or conversely to refer to the procurement of 'full' or 'unencumbered' vacant possession, thus imposing (by express qualification) a harsher and more robust standard of vacant possession (i.e. less likely to be subject to the usual *de minimis* threshold as established by case law).

The break right may provide, for example:

Option to Terminate
The Tenant may terminate this lease on the Break Date by giving not less than 6 months' prior written notice to the landlord, provided that on the Break Date the Tenant:
(1) *has paid all rents reserved by the Lease; and*
(2) *gives vacant possession to the Landlord.*

Practically speaking, it is advantageous for a tenant to avoid the procurement of vacant possession as a pre-condition for exercise of a break option. It would also be better if any agreement to provide vacant possession is kept entirely separate from the exercise of the break option. This way, the tenant will not be prevented from exercising the break if vacant possession cannot be given and the landlord, rather than seeking to contend that the lease is still continuing, will instead be forced to rely on other remedies to deal with the vacant possession issue (e.g. a damages claim following lease termination).[10] If vacant possession is an express precondition for successful operation of the break, such as above, then if vacant possession is not given on the break date, the purported exercise of the break right by a tenant will fail, and the lease will continue until its contractual expiry (or earlier termination in accordance with other provisions in the lease).

---

[9] See, for example, *Friar* v *Grey* [1850] 5 Ex 584.

[10] See chapter 8 for a discussion of damages.

It is important to note, however, that vacant possession can be a pre-condition for exercise of a break *impliedly*, by reference to a more general requirement on the break date. This is something often missed by tenants, with potentially devastating consequences.

IMPLIED BY COVENANTS

It is common for a lease to contain a more general pre-condition to exercise of the break option, namely that the tenant has 'complied' or 'materially complied' with all tenant covenants under the lease. For example, such a break right might provide:

> *Option to Terminate*
> *The Tenant may terminate this lease on the Break Date by giving not less than 6 months' prior written notice to the landlord, provided that up to and including the Break Date the Tenant has materially complied with all Tenant Covenants in the Lease.*

Where a break option is conditional upon the tenant's compliance, or material compliance, with covenants up to *and including* the break date, or at the break date, this will almost certainly encompass the 'yielding-up' obligation contained in the lease and which will take effect on termination.

Yielding-up clauses

A yielding-up obligation, as a tenant's covenant, is likely to provide:

> *Yielding-up*
> *At the expiration or sooner determination of the Lease, the Tenant is to yield-up the Premises to the Landlord in accordance with the covenants contained in the Lease.*

The theory behind the yielding-up obligation is that, upon lease termination, a landlord will want to have the premises returned in the state and condition in which they should be (pursuant to the covenants contained in the lease, which a tenant should have complied with). It is common for the yielding-up clause to refer to yielding-up 'the premises in repair', but 'it is considered that a covenant in this form takes effect as both a covenant to deliver up possession and also a covenant to deliver up in repair'.[11]

---

[11] Woodfall, *Landlord and Tenant* (Sweet & Maxwell, March 2010) 19.003.

According to *Henderson* v *Squire*,[12] there is an implied term in a lease that the tenant will deliver the demised premises back to the landlord at the end of the term in the absence of an express clause:

> *The question is, where there is a tenancy, and nothing is expressed as to delivering up possession at its determination, whether there is an implied contract that the tenant shall not only go out of possession, but restore the possession to the landlord...I think that there is such an implied contract.*[13]

This is subject to any express or contrary indications to the contrary.[14]

The express or implied obligation to yield-up (as a covenant by the tenant), taking effect on the break date (as the date of termination of the lease) will require the tenant to procure that any of the tenant's subtenants have vacated the demised premises, and for the tenant itself to vacate the premises (i.e. give vacant possession) – subject to rules relating to statutory protection.[15] As such, where a break is conditional on either yielding-up, or on compliance with the tenant's covenants at the break date (which will thereby incorporate into the break an express or implied covenant to yield-up), a form of vacant possession is likely to be an element of the pre-conditionality associated with successful operation of the break.

Precise wording of the clause

The actual wording of the yielding-up obligation will be fundamental. The normal phrase used in a yielding-up obligation is 'on the expiry or sooner determination of the term' but if there is reference in the yielding-up obligation only to 'expiry' (i.e. the end of the contractual term), then compliance may not be required on earlier termination pursuant to a break option. Further, if a break is conditional on compliance with tenant covenants *up to* the break date, and not 'up to and including' or 'at the break date', then the conditionality will not incorporate covenants (express or implied) which only take effect 'on' the break date itself. The yielding-up obligation (which includes the requirement to vacate) arises on the last day of the term (i.e. upon exercise of a break option, on the break date) but not before. As such, if the break is conditional on compliance with all tenants' covenants prior to the break

---

[12] *Henderson* v *Squire* [1869] LR 4 QB 170.

[13] ibid, per Blackburn J at 174. See also the judgment in *Harding* v *Crethorn* [1973] 1 Esp 56. See also Dowding & Reynolds *Dilapidations: The Modern Law and Practice* (London, Sweet & Maxwell, 4th Revised edition, 2008) para 14-06.

[14] See *Hyatt* v *Griffiths* [1981] 17 QB 505 and *Newson* v *Smythies* [1858] 1 F&F 477.

[15] *Henderson* [1869] LR 4 QB 170. For example, where a tenancy includes an absolute covenant to yield-up the property, this will not be applicable where the occupier has a statutory right to remain in possession of the property. The covenant to yield-up the property will be construed against the statutory background of the occupation of the tenant – see *Reynolds* v *Bannerman* [1922] 1 KB 719.

date, then the yielding-up covenant will not be one such covenant that the tenant must comply with on the break date in order to validly break the lease. In such a case, if vacant possession is not otherwise a pre-condition of the break, a tenant will not be required to give vacant possession in order to exercise the break itself.

Meaning of 'Yield-up'

In terms of how a tenant should 'yield-up', it has been said that:

> there has been very little guidance from case law as to what constitutes yielding-up...[n]either has Parliament ever prescribed the meaning of the expression.[16]

With that said, the 2004 decision in *John Laing Construction Ltd* v *Amber Pass Ltd*[17] provided some useful guidance.

In *John Laing*, the break right provided that the lease could be determined at the end of the 15th year of the term by serving six months notice in writing and paying 12 months' rent. The break clause also provided that upon the expiry of the notice, the tenant was required to:

> yield-up the entirety of the demised premises.[18]

The Court was asked to determine whether the break had been effectively exercised and the property had been yielded-up with vacant possession. The Court in this case decided that the tenant *had* effectively yielded-up the property, even though the tenant had failed to hand the keys back to the landlord, had left security personnel at the property and had not taken away removable concrete barriers, all of which (it was contended) prevented the landlord retaking possession of the property. The judgment made clear that a tenant does not have to undertake any specific acts to 'yield-up', but rather to manifestly demonstrate its wish to terminate the lease, enabling the landlord to be able to reoccupy. The task of the Court was:

> to look objectively at what had occurred and determine whether a clear intention had been manifested by the person whose acts were said to have brought about a termination to effect such termination, and whether the landlord could, if it wanted to, occupy the premises without difficulty or objection.[19]

---

[16] See Higgs, R 'Leave Your Keys on Your Way Out' (2005) 155 *New Law Journal* 149 in which the difficulties of appreciating what yielding-up at the end of the lease may involve were discussed. See also Martin, J 'Tenant's Break Options' (2003) 153 *New Law Journal* 759.

[17] *John Laing Construction Ltd* v *Amber Pass Ltd* [2004] 17 EGRL 128.

[18] ibid.

[19] ibid, per Robert Hildyard QC at 131.

In interpreting the decision, it would seem important to note that the security measures in question (including freestanding fencing and concrete barriers) were intended to protect the property from vandalism. The tenant had also arranged for 24 hour security specifically at the insistence of the landlord. The security arrangements being held *not* to constitute a hindrance that prevented the landlord retaking occupation of the property[20] can be seen to have been a finding of fact made in the specific context of the clear security problems that were apparent at the property, and the tenant's reasonable conduct in seeking to address these. This does, however, confirm again that any determination as to vacant possession will be a question of fact and degree, with reference to the specific circumstances and context of the case in issue.

Scope of yielding-up obligation

Whilst it has been held that the same principles from the freehold context apply to a tenant's obligation to deliver vacant possession in the leasehold scenario,[21] it has been noted that the yielding-up obligation in a lease may have a slightly reduced scope:

> *It is considered that a tenant who is obliged to deliver possession on the termination of the tenancy is to clear the property physically of chattels and other things which were not on the land at the beginning of the tenancy and have not subsequently become part of the land,* but it is doubtful whether his duty to yield-up possession extends to the removal of legal restrictions on the use of the property.[22]

This suggests that the obligation, in the context of yielding-up, may therefore be less extensive in scope.

Indeed, in the freehold context, where an *express obligation* to give vacant possession is engaged, the presence of a protected tenant *will* cause there to be a breach of the obligation on completion.[23] However, where the obligation on the party (in a leasehold context) is only to 'yield-up' the premises (which includes the return of possession of the premises), rather than to give vacant possession expressly, then the yielding-up obligation will likely *not* encompass an occupier who has a statutory right to remain in possession of the property; the covenant to yield-up the property will be construed against the statutory background of the occupation of the tenant.[24]

---

[20] See also *Jones* v *Merton LBC* [2008] 4 All ER 287, approving the decision in *John Laing*. The tenant retaining the keys to the flat was, in the circumstances, held not to be significant.

[21] In this regard see also *Legal & General Assurance Society* v *Expeditors (UK)* [2007] 1 P&CR 103.

[22] Woodfall, *Landlord and Tenant* (2010) 19.003. Emphasis added.

[23] See *Beard* v *Porter* [1948] 1 KB 321.

[24] See *Reynolds* v *Bannerman* [1922] 1 KB 719 and Woodfall, *Landlord and Tenant* (2010) 19.003. This rule applies even where the sub-tenancy was originally unlawful, if the unlawfulness has been waived by the landlord — see also *Watson* v *Saunders-Roe* [1947] KB 437.

Further, if the tenant is only required to 'reasonably comply' with all tenants' covenants at the break date (including the covenant to yield-up), the conditionality may be seen to be so qualified by such wording, unlike in cases where there is an express pre-condition to give vacant possession (as an absolute condition in its own right). If the tenant is only required to make a 'reasonable attempt' to yield-up (a lesser obligation still), it can be argued that a more generous *de minimis* threshold would therefore be applied by the courts to any impediments remaining in the premises at the break date.

Regardless of the scope of the yielding-up obligation in the leasehold context, it is clear that the implied requirement to give vacant possession can arise, even if it is qualified to some extent, as a consequence of the yielding-up obligation in the lease being incorporated as a condition of the break. This can be by virtue of the requirement that the tenant must have complied with all covenants up to and including the break date, which will include an express (or implied) covenant to yield-up the premises, or because yielding-up the property is itself a pre-condition for exercise of the break (as in the *John Laing* decision). It is common for tenants not to appreciate that vacant possession may impliedly be a pre-condition for exercise of the break in this way. When the issue is raised, normally *after* the break date, the tenant's opportunity to give vacant possession will have passed.

LANDLORD TACTICS ON CONDITIONAL BREAKS

In a changing, and recently downward, property market, it is common for tenants to seek to reduce operating overheads and source alternative (and normally smaller) premises. This can only be achieved if their current lease is brought to an end. Giving vacant possession is therefore usually essential for the tenant in order to ensure that its current lease does not continue past the break date.

If a tenant wishes to operate a break, then a landlord will need to consider, with reference to the precise wording of the break option, whether it is actually prepared to sit back and watch the tenant break the lease and leave. Indeed, a landlord may wish to frustrate the break by claiming, for example, non-compliance with covenants to repair under the lease, or payment of all rents reserved.[25] In similar terms, a landlord may choose to use the issue of vacant possession to prevent the tenant exercising a contractual break option in a lease if it would prefer the lease to continue.

A landlord may therefore wish to seek to prevent a tenant successfully exercising a break option in such circumstances, and vacant possession can be used as perhaps

---

[25] Martin, J 'Tenant's Break Options' (2003) 153 *New Law Journal* 759 explains that common pre-conditions to exercise a tenant's break option include, along with vacant possession, payment of all rent(s) due up to and including the break date. This itself can be an issue when the break date falls on a quarter day, upon which the whole quarter's rent falls due in advance.

the most effective and decisive means of preventing a lease coming to an end as per the tenant's intention. If the lease is expressly or impliedly conditional on vacant possession, the landlord may seek to construct an argument to the effect that vacant possession was not given at the material time (whether this is because of a physical or legal impediment preventing enjoyment of the property at the relevant date). If successful, this will thereby force a tenant to remain principally liable under the terms of the lease with the corresponding ongoing liabilities (including, most noticeably, the payment of rent).

In this regard, a landlord should be mindful of the following considerations:

1. What are the current market conditions relevant to the property, and the likely open market rent — how does this compare to the current rent passing under the lease?
2. What is the financial and covenant strength of the current tenant — are obligations of the tenant backed up with guarantees from other companies or individuals?
3. A landlord is not obliged to help the tenant understand or meet its break obligations.
4. Breaks are strictly enforced by the courts (no 'reasonableness' will be inferred). It is not necessarily the case that the tenant will be able to meet its break obligations, and terminate the lease as it is purporting to do.
5. Consider how helpful a landlord may wish to be in respect of the break. Potentially, consider refusing to agree any tenant's works proposals or classification of items as fixtures or chattels, simply reminding the tenant of its need to strictly comply with all conditionality associated with the break.
6. Do not start marketing the premises for a new tenant or put up a board before the break date. That is, do not intimate that the lease will come to an end as the tenant desires. This must obviously be balanced, however, with the need to ensure that a new tenant can be found in the event that the lease does come to an end pursuant to the break right.
7. Take early advice on the conditionality associated with the break and, even if a new tenant can be found or the landlord is prepared to allow the tenant to exercise the break, potentially seek to exploit any ambiguity associated with the conditionality of the break to maximise any financial settlement. For example, a tenant may be prepared to offer to pay a 'surrender premium' for the certainty of ensuring that the lease comes to an end as desired. Such a sum would likely incorporate other amounts falling due on lease termination (e.g. dilapidations payments and other terminal claims).

TENANT TACTICS ON CONDITIONAL BREAKS

A tenant should pay close attention to all pre-conditions associated with a break right in order to ensure that it complies strictly with all relevant requirements. Further, as far as possible, it should also seek to evidence such compliance in order to protect its position.

Common considerations for the tenant will include:

1. When agreeing terms for a new lease, avoid vacant possession as a break pre-condition, and make vacant possession an obligation that arises *following* the lease being broken.[26]
2. If the break is potentially difficult to operate due to conditionality which is ambiguous or onerous, take advice before serving the notice. This may include undertaking a compliance audit to identify any potential issues with procuring vacant possession. It is essential to understand obligations and prepare for complying with them. It will be sensible to appoint specialists prior to service of the notice.
3. Having served a notice, seek to engage the landlord from an early stage and seek confirmation of the steps the tenant needs to take in order to comply with any conditions. For example, seek to agree the status of items as fixtures and chattels and other requirements relevant to giving vacant possession or yielding-up the property on the break date.
4. If the tenant is concerned that the break may fail due to non-compliance with onerous conditionality, or that the costs of vacating and complying with other obligations may be excessive, consider asking the landlord to accept the break upon payment of an agreed amount as liquidated damages in relation to all liabilities under the lease, or to agree a surrender of the lease with the payment of an associated premium.
5. If the break is to be operated and complied with, ensure that such compliance is evidenced in order to avoid disputes later down the line. In vacant possession terms this will include taking pictures evidencing the physical state and condition of the property at the break date.
6. Make sure keys are returned before or on the break date and that the landlord can gain access (e.g. code/security passes) in order to avoid the landlord claiming that, whilst the premises have been left vacant, the landlord is unable to re-take possession without difficulty or objection.

More generally, the tenant should also be very careful before purporting to break a lease if it does not really want to move to alternative premises. In a tenant-friendly market, the tenant may threaten to exercise a break right or actually exercise it, with the intention of persuading the landlord to give the tenant more favourable terms moving forward. A tenant must bear in mind that once a break notice has been served, it cannot be withdrawn unilaterally, with any mutual waiver of the notice

---

[26] See later in respect of the Code for Leasing Business Premises in England and Wales, 2007, and the requirement to 'give up' rather than 'give vacant possession'.

being deemed to constitute the grant of a new lease with effect from the date of expiration of the break notice (subject to the break failing for non-compliance with conditionality, thus causing the lease to continue).

CONSEQUENCES OF NON COMPLIANCE FOR TENANT

As noted above, a break clause will be strictly construed and any conditions attached to the right must be strictly performed,[27] with time being of the essence in respect of the prescribed time limits in the clause.[28] Whether the relevant time for compliance is the date of service of the break notice, or the break date (or in some cases, both), is a matter of construction of the break clause. For example, sometimes rent may have to be paid up-to-date to serve the notice, and then to operate the break at the break date itself. With respect to vacant possession, this is an obligation for which compliance can only be required on the break date itself (and not before).

If the tenant fails to operate the break option successfully and remains tied to its current lease, along with a new tenancy that may already have commenced, or the tenant has become contractually bound to enter into, this could lead to dire financial circumstances. It is common for tenants to misunderstand the requirements for successful operation of a break option and unwittingly find that they remain liable and tied to the covenants contained in their current lease because of vacant possession not being given on completion.

If a tenant fails to operate a break right due to non-compliance with conditionality, a tenant may only be able to divest itself of its ongoing liability by assigning or subletting the lease to a third party (but this will be determined by the terms of the lease and relevant applicable statute).[29] The risks for a tenant of not giving vacant possession can be very high therefore.

VACANT POSSESSION ISSUES ON CONDITIONAL BREAKS FOR SURVEYORS

As noted in chapter 7, the state and condition of a given premises may well be relevant to the procurement of vacant possession, and therefore a matter which must be considered carefully upon the exercise of breaks which are conditional on either vacant possession, yielding-up, or repair obligations more generally.

The state and condition that the premises must be returned in will be determined by the terms of the lease (including, in particular, obligations relating to reinstatement) and also supplemental licences and documents pertaining to alterations or

---

[27] *Reed Personal Services* v *American Express* [1997] 1 EGLR 229.

[28] *United Scientific* v *Burnley Corporation* [1978] AC 904.

[29] For example, the statutory provisions of the Landlord and Tenant Act 1927.

other changes. The terms of these will be crucial in seeking to establish whether certain 'items' in the premises can stay, or must go.

The cost of reinstating, or stripping out, a property and complying with repairing obligations under the lease, and then making good any damage caused by the removal of structures (or equivalent items) from the premises, can be excessive. The extent of a tenant's obligations will commonly be a source of dispute, and uncertainty. A tenant will want to spend the minimum possible to comply with its obligation to give vacant possession (and any other tenants' covenants upon which the break is conditional), but at the same time will not want to fall foul of the conditionality, thus causing the lease to continue.

A yielding-up clause will often expressly require the tenant to remove fixtures and any alterations it has made to the property, in addition to making good damage caused in connection with their removal. Such an express provision is common in yielding-up clauses to avoid the issue of reinstatement from being overlooked in the repairing covenants or in any licences for alterations. When such a clause is included it will also be necessary to consider the extent of its scope. For example, such a clause is likely to catch any alterations that have been carried out in breach of covenant along with alterations for which consent was not required.[30]

So-called 'strip-out and reinstatement obligations', quite commonly, will not be included in a yielding-up clause in the lease. This is because a reinstatement obligation may be viewed as onerous at rent review if the reinstatement obligation is not specifically referred to as a 'disregard'. Further, a landlord may not actually want the tenant to reinstate if, for example, the property is to be demolished, and may not want such a default provision to be provided for by the lease therefore.

Further, a particular issue relates to works that are carried out in order to comply with statute. There can often be a conflict between, on the one hand, a requirement of statute that certain works be undertaken during the life of a lease (remaining in place thereafter) and, on the other, a reinstatement obligation in a yielding-up clause requiring the removal of such works.

This position can be further complicated if no reinstatement provision is included at all. If the lease does not contain any provisions dealing with reinstatement, the tenant's obligations will be entirely determined by the terms of the lease and any consents granted for alterations, or individual enforcement measures.

A number of general principles can be extracted from the applicable case law in relation to these issues:

1. The tenant is not generally obliged to remove things that are 'part of the building' or that are landlord's fixtures.

---

[30] A covenant against alterations may not apply to minor works — see *Bickmore* v *Dimmer* [1903] 1 Ch 158.

2. The tenant is not generally *obliged* to remove tenant's fixtures, although it is entitled to do so until the last minute of the term.[31] A landlord who does not want a tenant to remove such fixtures should make this clear in the lease or licence for alterations.
3. In relation to any dispute about the meaning of any express covenant to remove fixtures, the covenant will be interpreted using the normal principles governing interpretation of contracts and other legal documents.
4. A covenant to yield-up the property in good repair and condition may require the tenant to remove redundant equipment, something often not appreciated.[32]

One major issue which often raises its head concerns partitioning. Often, leases state that tenants must remove demountable partitioning, but in the absence of clear words, removal of all partitioning may be required (or advisable). A covenant to leave premises 'in good repair and condition' was held to require the tenant to remove partitioning in a case from the Australian courts, but it is unclear to what extent this will be followed in England and Wales.[33]

One way to deal with these commonly encountered issues is to make any reinstatement obligations in the lease conditional on the landlord giving the tenant notice that it does want fixtures and alterations removed, and the premises reinstated. This is advantageous to the tenant because, in addition to the certainty that such a procedure would provide, removal and reinstatement is expensive and time-consuming. In turn, reinstatement is a waste of time if the landlord intends to demolish the premises and a landlord may rather agree an enhanced dilapidations settlement instead of insisting on the unnecessary works being undertaken. Further, the landlord may be more able to secure a re-letting of the property to a new tenant with the alterations and fixtures in place, and as such may prefer that they remain in the premises, regardless of whether there is an express or implied reinstatement obligation to remove these on lease termination.

Clearly, the state and condition of a given premises can be relevant to the procurement of vacant possession, and therefore a matter which must be considered carefully upon the exercise of breaks which are conditional on either vacant possession, yielding-up, or repair obligations more generally. The practical task of compliance will normally fall to the building surveyor.

CODE FOR LEASING BUSINESS PREMISES

It is correct to say that the property industry has shown awareness of the particular problem posed by the issue of vacant possession on the exercise of conditional

---

[31] *Never-stop Railway (Wembley) Ltd* v *British Empire Exhibition (1924) Inc* [1926] Ch 877.
[32] *Shortlands Investment Ltd* v *Cargill* [1995] 1 EGLR 51.
[33] See *Wincant Pty Ltd* v *State of South Australia* [1997] 69 SASR 126.

breaks. So much so, that the Code for Leasing Business Premises in England and Wales 2007[34] suggests that vacant possession should be avoided as a pre-condition of a break altogether.

The Code is the result of collaboration between commercial property professionals and industry bodies representing both owners (landlords) and occupiers (tenants). The Code aims to promote fairness in commercial leases, and recognises a need to increase awareness of property issues, especially among small businesses, ensuring that occupiers of business premises have the information necessary to negotiate the best deal available to them. It is commonly viewed as being shorter, easier to understand and more focused than the previous version of the Code in 2004. The Code is entirely voluntary and there is no requirement to comply with every single part of the Code. However, the working group that re-wrote the Code sought to achieve a workable document that provides a fair and level playing field between the parties. In respect of break options, the Code prescribes only three pre-conditions, if any, to the exercise of a tenant's break option:

1. The tenant should be up-to-date with the main rent, (the basic rent, not service charge).
2. The tenant should give up occupation. This is *not* the same as giving vacant possession — see below.
3. The tenant should leave behind no continuing subleases.

What is important to note about these three conditions is that they are wholly within the tenant's ability to comply with.

As noted above, in *John Laing*,[35] the High Court held that a tenant had effectively exercised a break clause and had yielded-up the property with vacant possession, although it had left behind certain security measures intended to protect the property from vandalism. The Court held first that the tenant had clearly demonstrated its wish to terminate the lease and that it was the Court's task to assess whether such an intention to determine the lease had been manifested by its conduct. It is this sense of seeking to 'give up' the premises that the Code is seeking to incorporate.

The second part of the *John Laing* test, as to whether, pursuant to the intended determination by the tenant, the landlord could, if it wanted to, 'reoccupy the property without difficulty or objection' is not necessary under the proposed conditionality of the Code. This is because the requirement to 'give up' occupation

---

[34] Code for Leasing Business Premises in England and Wales 2007, p 1. Copy available from www.leasingbusinesspremises.co.uk.

[35] *John Laing* [2004] 17 EG 128.

does not, as commonly misunderstood, require the party to give vacant possession. As the Code states:

> *any disputes about what has been left behind or removed should be settled later (like with normal lease expiry).*[36]

As such, the Code avoids some of the commonest issues relating to the giving of vacant possession which often catch the tenant out when seeking to exercise a conditional break option. The effect of adopting the Code is thus to take from the landlord the ability to seek to frustrate a tenant's compliance on the grounds of the non-procurement of vacant possession.[37]

As the following table explains, whilst the test for vacant possession under the decisions in *Cumberland Consolidated Holdings Ltd* v *Ireland*[38] and *John Laing* comprise two limbs, the Code only therefore requires the first limb to be complied with on the exercise of a tenant's break option.

Table 9.1 A Comparison between the Requirements in *Cumberland* and *John Laing*, and the Code for Leasing Business Premises in England and Wales 2007

	Cumberland	John Laing	Comment	Lease Code
Limb 1	Does the conduct of party required to give vacant possession suggest that possession is being passed?	Do acts of the party required to give vacant possession manifest a desire to give/return possession?	Directed at the intention of the party required to give vacant possession, as demonstrated by their actions.	'Give up' possession – intend to give possession back to the landlord.
Limb 2	Does the impediment (objectively speaking) substantially prevent or interfere with possession of the property or a substantial part?	Is it possible (objectively speaking) for the party with the right to vacant possession to (re)occupy without difficulty or objection?	Focusing on whether the party with the right to vacant possession is actually able (objectively speaking) to occupy the property in question.	**NOT REQUIRED** 'Any disputes about what has been left behind or removed should be settled later'. This should also extend to other impediments to vacant possession.

Of course, if vacant possession must be given upon the lease being determined (even though not a pre-condition for exercise of the break option), the landlord will be able to advance a damages claim in respect of the tenant's failure to comply with

---

[36] Code for Leasing Business Premises in England and Wales 2007, p 1.

[37] See Shaw, K. 'Bone of Contention' (2009) *Estates Gazette* 58–59.

[38] *Cumberland Consolidated Holdings Ltd* v *Ireland* [1946] KB 264.

the relevant covenant — usually to yield-up in a particular state and condition — although this will *not* prevent the lease from determining.[39]

Whatever one's view of the requirement to give vacant possession, an obligation commonly misunderstood and incorrectly interpreted, it is clear that the draftsmen of the Code do not consider it appropriate to make the provision of vacant possession (in its traditionally understood sense) a pre-condition to the exercise of a tenant's break option, with the tenant only required to 'give up' occupation (i.e. only the first limb of the *Cumberland* and *John Laing* tests). Whilst obviously the bargaining strength and specific circumstances of any lease negotiations will be key in settling the terms for the exercise of a break option, this reflects a clear policy of seeking to obviate the problems associated with the obligation by simply removing, as a pre-condition altogether, the troublesome element of vacant possession (namely the second limb of the *Cumberland* and *John Laing* tests, which a tenant may have difficulty complying with). This is perhaps an example of the industry seeking to redress the balance for the tenant who, under the Code, can avoid having to give vacant possession when exercising a break; a position markedly different to the strict legal requirement of being required to give vacant possession where 'vacant possession proper' is an express pre-condition for exercise of the break, as is commonly found in non-code compliant leases and tenancies.[40]

Vacant possession on rent review

A particular instance in which vacant possession will be relevant in the leasehold context is with respect to the assumed state of occupation of a given premises for the purposes of a rent review valuation. Whether the premises (or part thereof) are deemed to be occupied for the purposes of such a calculation can be material to the reviewed rental value.

This section, with reference to the applicable statute, case law and rent review hypotheses, explains the assumptions that can be made as to a premises being valued with vacant possession, along with the practical consequences to both landlord and tenant.

---

[39] On normal lease termination a tenant is liable to pay compensation for use and occupation for any period in which the landlord is unable to retake possession. An equivalent amount may be sought as damages for breach of the implied obligation to deliver possession (*Henderson* [1869] 4 LR 4 QB 170) or an express covenant to yield-up (*Thames Manufacturing Co Ltd* v *Perrotts (Nichol & Peyton) Ltd* [1984] 50 P&CR 1). A landlord may also seek to recover from a tenant its costs incurred in taking proceedings to so recover possession.

[40] See Shaw, K. 'Bone of Contention' (2009) *Estates Gazette* 58–59.

Rent Review Generally

Some leases are granted in return for a premium, or capital sum of money. More commonly in cases of commercial leases, the rent payable will be the market rent (also known as rack rent). A commercial rent is usually an annual sum, payable in advance in four equal instalments, often on the usual quarter days.

Most leases of commercial premises will also include a provision which allows the rent reserved to change over time. The effect of this is to ensure that the landlord continues to receive a rent which fairly reflects the real value of the let property. Long leases, which were granted at a premium, may also provide for the rent to increase over time, even though the change is not likely to be as linked to the market value of the property as with a commercial lease. The Court of Appeal in *Equity & Law Life Assurance Society Plc v Bodfield Ltd*[41] took the view that the purpose of a rent review clause was to allow the rent to keep track with inflation and changes in the value of property.

A standard rent review clause is likely to include a number of 'assumptions' and 'disregards'. A clause stating that a certain fact is to be assumed to exist will be an assumption. It is common, for example, for a rent review provision to provide that a valuation of the property at a rent review should proceed on the basis that the tenant has complied with all of its obligations and covenants under the lease. By contrast, a disregard is a clause stating that a certain fact is not to be taken into account when determining the rental value of a property. Works carried out by the tenant, other than those which a tenant may be obliged to undertake in compliance with its covenants under the lease, would be one example.

Statute and Hypothesis

Section 34 of the Landlord and Tenant Act 1954 directs the court to disregard (*inter alia*) certain matters when determining the rent that is to be payable for a new business tenancy. The disregards contained in this section are often referred to in standard rent review clauses in leases.

In respect of vacant possession, two disregards are of importance, namely:

- any effect on rent of the fact that the tenant has or its predecessors in title have been in occupation of the holding; and
- any goodwill attached to the holding by reason of the carrying on thereat of the business of the tenant (whether by the tenant or by a predecessor of the tenant in that business).[42]

---

[41] *Equity & Law Life Assurance Society Plc v Bodfield Ltd* [1987] 1 EGLR 124.

[42] Landlord and Tenant Act 1954 s 34(1)(a) and (b).

The rent review hypothesis assumes, as at the review date, that the lease including the rent review clause *does not exist*. Logically, any occupation by a tenant in possession has to be disregarded therefore, and it is to be assumed that the tenant is not in occupation. In *FR Evans (Leeds)* v *English Electric*,[43] the valuer was instructed to determine the rent at which the premises were 'worth to be let with vacant possession on the open market…disregarding…any effect on rent of the fact that the lessee or its tenant has been in occupation of the demised premises'. In that case, Donaldson J remarked that:

> *It is implicit in this instruction, and is expressed in [the disregard] that the tenants are to be deemed to have moved out or never to have occupied the premises.*[44]

With that said, it does not follow automatically that it can therefore be assumed that there is vacant possession of the premises subject to the rent review. This is because, whilst not occupied by the tenant, there may be other third parties in occupation at the date of the rent review (for example, subtenants of the actual tenant or those occupying under a licence arrangement).

Indeed, where the rent review clause does not, expressly or impliedly, direct an assumption of vacant possession, but does contain a disregard of the occupation of the tenant, the question can arise as to whether this disregard is intended to extend to any subtenants or licensees of the tenant that may be occupying the premises (or whether it is simply a reference on to the actual tenant itself). For example, in *Daejan Investments* v *Cornwall Coast Country Club*,[45] the disregard was held to apply only to the tenant, and the existence of the subtenant/licensee was to be taken into account for the purposes of the valuation.

This issue goes to the heart of the question of whether there can be an assumption of vacant possession on rent review.

ASSUMPTION OF A VALUATION WITH VACANT POSSESSION

It is common for a rent review clause to expressly provide that the valuation shall assume that the premises are to be let with vacant possession of the whole. With that said, a number of clauses remain silent on this point. In such a case, it will be necessary to seek to determine whether the lease and the rent review clause

---

[43] *FR Evans (Leeds)* v *English Electric* [1978] 1 EGLR 93.
[44] ibid, per Donaldson J at 94. See later in connection with the suggestion that the tenants 'never occupied'.
[45] *Daejan Investments* v *Cornwall Coast Country Club* [1985] 1 EGLR 77.

'implicitly contemplated' that vacant possession of the premises was to be the basis upon which the rent review should take effect.

According to *Hill Samuel Life Assurance* v *Preston Borough Council,*[46] when assessing a rack rental lease it is likely that a valuation with vacant possession is contemplated, but this will be subject to express contrary indications as manifest in the wording of the lease more generally. An example of a tenant seeking to rely on express wording to the contrary can be found in the decision in *Priorflow* v *TR Services.*[47] Here, the tenant argued that the lease should be valued subject to underleases, rather than with vacant possession, upon the 'presumption of reality' (which contends that the hypothetical lease should reflect the real one as closely as possible).[48] The tenant also relied on the definition of 'Demised Premises' which, it was claimed, imported into the definition the existence of the underleases, as well as the physical ambit of the premises. Whilst the tenant therefore sought to rely on the express lease wording to rebut the assumption of a valuation with vacant possession, the Court in the case did not consider that the language of the lease was 'sufficiently clear and unambiguous' to override any such contentions and accordingly it was held that vacant possession could be assumed for the purposes of the valuation.

In terms of seeking to contend that it is reasonable to claim that it *was* intended that the premises would be valued on rent review with vacant possession, it is relevant to consider two factors in particular:

1. Were the premises (in whole or part) already sublet when the lease was originally granted?
2. Did the original letting take effect in circumstances where a subletting of part may have been contemplated?

---

[46] *Hill Samuel Life Assurance* v *Preston Borough Council* [1990] 36 EG 111.

[47] *Priorflow* v *TR Services* [1992] Unreported. Discussed in Reynolds K and Fetherstonhaugh, G *Handbook of Rent Review* (Sweet & Maxwell, 2009) p 612.

[48] See *Basingstoke and Deane Borough Council* v *Secretary of State for Communities and Local Government* [2009] EWHC 1012 (Admin) in which Nicholls LJ explained the principle in these terms: 'Of course rent review clauses may, and often do, require a valuer to make his valuation on a basis which departs in one or more respects from the subsisting terms of the actual existing lease. But if and in so far as a rent review clause does not so require, either expressly or by necessary implication, it seems to us that in general, and subject to a special context indicating otherwise in a particular case, the parties are to be taken as having intended that the notional letting postulated by their rent review clause is to be a letting on the same terms (other than as to quantum of rent) as those still subsisting between the parties in the actual existing lease. The parties are to be taken as having so intended, because that would accord with, and give effect to, the general intention underlying the incorporation by them of a rent review clause into their lease.'

An example of the application of these principles comes from the decision in *Laura Investment Co* v *Havering London Borough Council*,[49] where undeveloped land was demised on a lease which contained a covenant by the tenant to divide the land into plots to be under-let for the construction of buildings. At the first rent review, the tenant had divided up and underlet the premises as the lease had required, with the erection of buildings already having taken place. The Court held that the rent was to be assessed on the assumption that the premises were subject to the underleases in existence at the date of the rent review, rather than with vacant possession, ostensibly because sublettings of part were clearly contemplated by the lease. In *Scottish & Newcastle Breweries Ltd* v *Sir Richard Sutton's Settled Estates*,[50] an express disregard on rent of the occupation of the tenant or any persons deriving title under it was contained in the lease. The tenant, however, asserted that that the lease should be valued subject to the underlettings existing on the review date. Whilst the Court noted that an assumption of vacant possession was generally likely to be the parties' intention, subject to express contrary indications, on a proper interpretation of the lease it was held that the notional letting was to be assumed to be subject to the underlettings existing on the review date because (*inter alia*) the sublettings had been contemplated and agreed to by the parties from the outset.

Such decisions can, however, be contrasted with *Leigh* v *Certibilt Investments*.[51] In this case, notwithstanding the demised premises were sublet in part at the date of the grant of the lease, and that it was contemplated that the premises would be further sublet, the Court held it appropriate to value the premises, for the purposes of the rent review, with vacant possession. Of importance here appeared to be the fact that the rent review clause required the determination of the underlying rack rental value of 'the estate', thus making a valuation with the assumption of vacant possession appropriate. This reflects how the wider context of any given case must be considered carefully.

Lawful subletting

A number of authorities deal with the implied assumption of vacant possession for the purposes of valuation on rent review, and the issue of *lawful* sublettings.

In *Oscroft* v *Benabo*,[52] the task of the Court of Appeal was to decide whether, for the purposes of section 34 of the Landlord and Tenant Act 1954, premises which were partially sublet at the date when the rent under the new tenancy comes to be

---

[49] *Laura Investment Co* v *Havering London Borough Council* [1993] 08 EG 120.

[50] *Scottish & Newcastle Breweries Ltd* v *Sir Richard Sutton's Settled Estates* [1985] 2 EGLR 130.

[51] *Leigh* v *Certibilt Investments* [1988] 1 EGLR 116.

[52] *Oscroft* v *Benabo* [1967] 2 All ER 548.

fixed by the Court, should be valued on the basis of vacant possession or part possession. In this regard, it must be noted that section 34 does not expressly direct a determination on the basis that the premises are to be let with vacant possession. Willmer LJ stated:

> *In fixing what a willing lessor would be prepared to take [the open market rent under Section 34], it seems to me that all the circumstances of the particular case, including the fact of any existing sub-tenancy, must necessarily be taken into consideration.*[53]

Such *obiter* comments clearly suggest that in cases where there has been a lawful subletting, the assumption of vacant possession cannot be implied as a matter of course, with all the circumstances of the particular case needing to be taken into account. In *Forte & Company* v *General Accident Life Assurance*,[54] the lease, granted subject to an existing underlease of part, included a disregard on the basis of section 34 of the Landlord and Tenant 1954 Act. The Court followed *Oscroft*, and made clear that it could not be assumed as part of the hypothesis of an open market letting that the existence of that underlease could simply be disregarded as not relevant.

The decision in *Forte* can be contrasted with the ruling in *Avon County Council* v *Alliance Property Co*,[55] where the Court held that there was to be an assumption of vacant possession notwithstanding that the premises had been entirely sublet at the date of the lease. This decision seems to have been made specifically on the basis that the premises in question were, under the terms of the lease, expressly required to be valued on review 'as a whole' and not on any other basis.[56]

Subletting in breach of covenant

Whilst *Oscroft* suggests that, in cases where there has been a lawful subletting, the assumption of vacant possession cannot be implied as a matter of course, where the subletting was in breach of a covenant in the lease, the converse position is appropriate. The *Handbook of Rent Review* explains that:

> *Where the subletting in question was in breach of covenant in the lease, and that breach has not been waived by the review date, the sub-letting should be ignored in deciding whether vacant possession should be assumed (and therefore be ignored for valuation purposes), if, as in most cases, the effect*

---

[53] ibid, per Willmer LJ at 553.

[54] *Forte & Company* v *General Accident Life Assurance* [1986] 2 EGLR 115.

[55] *Avon County Council* v *Alliance Property Co* [1981] 1 EGLR 110.

[56] Other cases, however, have also assumed the premises to be vacant on the date of rent review − see *Bishopsgate (99)* v *Prudential Assurance Co* [1985] 1 EGLR 72.

of taking into account the subletting would be to decrease the rent. To do otherwise would allow the tenant to benefit from his own wrong in a way that the parties cannot have intended when the lease was entered into.[57]

The *Handbook of Rent Review* correctly points out that such an argument is further supported in cases where there exists an express requirement that it should be assumed that the tenant has complied with its covenants under the lease.

Previous occupation

Whilst the rent review hypothesis assumes, as at the review date, that the lease including the rent review clause *does not exist* and that any tenant is not in occupation, the fact of a tenant's *previous* occupation can be relevant however. It was the view of Donaldson J in *FR Evans* that it was implicit in the assumption of vacant possession that 'the tenants are to be deemed to have moved out or never to have occupied the premises',[58] but the *Handbook of Rent Review* correctly questions whether the fact of possession of the tenant needing to be disregarded also requires that the tenant should be assumed *never* to have occupied the premises previously.[59] This issue can be overcome by express wording in the rent review clause which expressly disregards any effect on rent as a consequence of the tenant (or predecessors in title) having *ever been* in occupation. It is also possible to achieve this by incorporating the statutory disregard in section 34(1)(a) of the Landlord and Tenant Act 1954 which achieves the same objective in similar terms. In the absence of such provision, the previous occupation of a tenant may be a relevant factor.

JUSTIFICATION AND CONSEQUENCES OF ASSUMPTION

When a landlord and a prospective tenant (who is already in occupation), are negotiating a new rent, the fact that the tenant is already in occupation could potentially induce the tenant to offer a sum greater than it would offer if it were not in occupation of the demised premises that are subject to the review. This is logical when one considers that the tenant in occupation is likely to view the avoidance of the expense and disruption to its business of having to vacate and relocate to other premises as having an intrinsically greater value, which would not be apparent for a third party tenant moving to the premises from elsewhere. Many confuse this with the notion of goodwill (which is discussed below), but this consideration would still be relevant in circumstances where the tenant had

---

[57] Reynolds and Fetherstonhaugh, *Handbook of Rent Review* (2009) p 612.
[58] *FR Evans* [1978] 1 EGLR 93, per Donaldson J at 94.
[59] Reynolds and Fetherstonhaugh, *Handbook of Rent Review* (2009) p 612.

not undertaken business from the premises or was a business to which goodwill was not a relevant matter. The tenant, in occupation, would still potentially attach a value to remaining in the current premises as compared to having to move elsewhere.

As such, the mischief behind the disregard of occupation, and assumption of vacant possession, is to prevent the landlord benefitting from an increase in the rent for the premises attributable to the fact of occupation by the tenant itself.[60]

In terms of the consequences of the assumption of a valuation of the premises with vacant possession, if vacant possession is assumed on rent review, the existence at the review date of any subletting must be treated as irrelevant. Conversely, if vacant possession is not to be assumed, then valuation becomes the central issue. The *Handbook of Rent Review* provides a useful example of how valuation can be an issue:

> *The fact that a small detached section of the property is sub-let may have virtually no effect on value. The fact that an upper part above a shop with no separate entrance is sub-let on a tenancy protected by the Rent Acts may have a startling effect on value. Where there has been a decline in the market since the subletting was created, the effect of valuing subject to (and with the benefit) of the sublease may be to increase the rent.[61]*

Further, if vacant possession of the premises is to be assumed, or there is, as a matter of fact, no tenant in possession, it follows that all tenant's fixtures and chattels must be assumed to have been removed by the review date.[62] In such a case it will be necessary for a valuer to identify which items are removable (i.e. which are tenant's fixtures and chattels) and determine whether an incoming tenant would need to replace these items in order to make beneficial use of the premises. If such items would need to be replaced, then the likely cost to the incoming tenant in so doing would need to be ascertained and reflected in any likely discount that the incoming tenant would seek, to reflect such cost, from the likely passing rent. The reduction can, in some cases, be very small, but in others, rather substantial. The implications are most likely to be the greatest in cases where the property was demised as a shell building and the majority of the fixtures contained therein belong to the tenant.[63] For this reason, rent review clauses may also assume that the

---

[60] See also *East Coast Amusement Co* v *British Transport Board* [1965] AC 58 and *Hambros Bank Executor & Trustee Co* v *Superdrug Stores* [1985] 1 EGLR 99.

[61] Reynolds and Fetherstonhaugh, *Handbook of Rent Review* (2009) p 612.

[62] *New Zealand Government Property Corporation* v *HM & S* [1982] Q.B. 1145 and Daejan[1985] 1 EGLR 77.

[63] See *GREA Real Property Investments* v *Williams* [1979] 1 EGLR 121 and *Estates Projects* v *London Borough of Greenwich* [1979] 2 EGLR 85.

premises are fit for occupation by the hypothetical tenant, to avoid needing to factor into any valuation the need for a rent free period to allow for fit out works to be undertaken.[64]

It is important to note that, even in cases where the lease stipulates that the tenant's fixtures are to become the property of the landlord at the end of the term, the effect of the disregard of improvements may still cause such fixtures to be disregarded as tenant's improvements on rent review. In *Bretair Pty Ltd* v *Lenro Properties Pty Ltd*,[65] the lease stated that:

> *When the Term comes to an end the Lessee's improvements, fixtures and fittings and all property affixed to the premises shall become the Lessor's property.*[66]

This was subject to a few limited exceptions. The landlord argued that the revised rent should be based upon the value of the premises including those fixtures, but it was held by the tribunal that the fixtures should be disregarded. This was by virtue of the stipulation in the rent review clause that the valuer should 'ignore any fixtures and fittings which the lessee has the right to remove from the premises and improvements voluntarily made by the lessee' (i.e. before they become the property of the landlord upon expiry of the term).

GOODWILL

Rent review clauses usually require (either expressly, or by incorporating the statutory disregard under section 34(1)(b) of the Landlord and Tenant Act 1954) that the following should be disregarded:

> *any goodwill attached to the holding by reason of the carrying on thereat of the business of the tenant (whether by the tenant or by a predecessor of the tenant in that business).*[67]

In *IRC* v *Midler & Co's Margarine Ltd*,[68] Lord Macnaghten defined goodwill as 'the expectation that an existing customer clientele of a business will continue'.[69]

The disregard of goodwill is tied to the business and not to the tenancy. Accordingly, the goodwill does not have to have been built up during the current

---

[64] See Bamford, K. *Amending a Commercial Lease* (Wiltshire, Bloomsbury Professional Ltd, 2010) p 184.

[65] *Bretair Pty Ltd* v *Lenro Properties Pty Ltd* [2004] VCAT 1141.

[66] ibid at 1148.

[67] Landlord and Tenant Act 1954 s 34(1)(b).

[68] *IRC* v *Midler & Co's Margarine Ltd* [1901] AC 217.

[69] ibid, per Lord Macnaghten at 219.

tenancy (or indeed during any tenancy), but rather reflects continuity of business. It is therefore to be expected that a disregard of the tenant's occupation (which, as noted above, is technically separate to goodwill) is likely to be accompanied by a disregard of goodwill in its own right. In *Prudential Assurance Co v Grand Metropolitan Estates,*[70] the lease contained disregards very similar to those contained in section 34(1)(a) of the Landlord and Tenant Act 1954, but did not include any express reference to the disregard of goodwill, as is found in those statutory provisions. The judge was, however, persuaded that in order to disregard the tenant's occupation, it would be necessary to disregard any goodwill generated by that occupation as well.

INTERPRETATION AND BEST PRACTICE

As demonstrated in this section, the law covering vacant possession on rent review has its basis in the law of contract and consequent principles of interpretation, with only a small amount of legislation being applicable.

In seeking to construe a rent review clause, Nicholls LJ explained in *Basingstoke and Deane Borough Council v The Host Group Ltd*[71] that the Court will seek:

> ...to identify and declare... the intention of the parties to the lease expressed in that clause. Thus, like all points of construction, the meaning of this rent review clause depends upon the particular language used interpreted having regard to the context provided by the whole document and the matrix of the material surrounding circumstances. While recognising, therefore, that the particular language used will always be of paramount importance, it is proper and only sensible, when construing a rent review clause, to have in mind what normally is the commercial purpose of such a clause.[72]

This is why the decisions referred to in this section are fact specific, in the context of certain overriding principles.

Because the wording of the rent review clause is so important, extra care should be taken when drafting rent review provisions, which need to be clear and unambiguous. Whilst a clause may seem obvious to understand when first drafted and in the early stages of negotiation, its actual meaning can later cause litigation when the rent review clause needs to be interpreted during the lease term when rent review comes round.

---

[70] *Prudential Assurance Co v Grand Metropolitan Estates* [1993] 2 EGLR 153; [1993] 32 EG 74.

[71] *Basingstoke and Deane Borough Council v The Host Group Ltd* [1987] 2 EGLR 147.

[72] ibid, per Nicholls LJ at 152.

Lawyers need to be especially careful about making bespoke amendments to rent review clauses to reflect any special features of a given property, and think through the consequences more generally to operation of the clause. It is also important to bear in mind the terms of a rent review clause when varying a lease, as a subsequent variation may change the interpretation of the rent review clause as was intended when the lease was originally drafted and negotiated.

With regard to vacant possession, a rent review clause should clearly provide that the property will be valued on the basis that it is being let with vacant possession and, as such, all occupation (and past occupation) of the tenant (and its predecessors in title) should be ignored. It should also expressly provide that all tenant's fixtures and chattels should be disregarded on valuation and that the premises are ready to be occupied by a hypothetical tenant for its required purposes. This has the effect of avoiding the need for any implications to arise.

If subleases (either present at the date of lease or to be granted thereafter) are intended to be part of any valuation on rent review, it is again better to expressly state this, or otherwise clearly record that the existence (or potential existence) of any sublease is to be ignored. This avoids the uncertainty as to whether only the occupation of the tenant is to disregarded on rent review.

Summary

In summary, rent review clauses will usually provide that the property will be valued on the basis that it is being let with vacant possession. As such, the tenant will be treated as having moved out (or as never having occupied the premises), and the rent will be calculated on the assumption that the tenant has removed its tenant's fixtures and chattels from the premises. Further, the premises may be assumed to be ready for occupation by a tenant for all necessary purposes (to avoid having to factor into a valuation a rent free period to allow for fit out works to be undertaken).

Even if there is no express provision in the lease stating that the premises are to be valued on the basis that they are to be let with vacant possession, the normal assumption is that a lease should be valued on such a basis (subject to express contrary wording or indications).

A particular issue relates to subleases. Whether the lease was granted specifically subject to an existing tenancy, or the lease clearly contemplated the creation of an underlease, will be the important factors to consider. Case law indicates that all relevant circumstances must be taken into account in such instances. A lawful sublease (or licence) may be taken into account in the rent review valuation in appropriate cases, with the consequent effect on valuation. If the sublease was granted in breach of covenant, however, then the subletting should be ignored in deciding whether vacant possession should be assumed; otherwise the effect of taking into account the subletting would likely be to decrease the rent, unfairly allowing the tenant to benefit from its own wrongdoing by acting in breach of covenant and granting a sublease (something clearly not contemplated by the lease).

Chapter 10

Case Studies

Chapter Outline

Case study 1: vacant possession in the freehold context

This case study covers a variety of vacant possession issues which may arise on the sale and purchase of freehold property.

Vacant Possession: Law and Practice. ISBN: 978-0-08-096680-9

Consider whether there has been a breach of the obligation to give vacant possession on the facts in the following scenarios, and what principles are relevant to the determination in each case.

Scenario 1

A seller contracts to convey land to a purchaser. Upon exchange the purchaser knew of large items in the property (it is not clear as to whether these can easily be removed or not).

The only possible reference to vacant possession in the contract is a clause stating that 'the property is sold free from legal incumbrances on completion'. It is unclear as to whether this refers to vacant possession expressly.

On completion, the purchaser claims that the seller has not given vacant possession because the large items remained on the property at completion. The purchaser claims that the seller expressly contracted to provide vacant possession and is therefore in breach.

Conversely, the seller claims that the clause in question referred only to title issues and therefore that any obligation to procure vacant possession is implied. The seller further claims that because the purchaser was aware of the large items in the property on exchange and that they are clearly *irremovable*, the implied obligation to give vacant possession did not extend to include these items and therefore that the seller is not in breach.

Analysis

This question addresses the issue of whether, and if so how, an obligation to give vacant possession has arisen. From a practitioners point of view, ascertaining *how* the obligation to give vacant possession has arisen (by express provision or implied under open contract rules) is essential in order to determine what the obligation encompasses.

It is unclear as to whether the clause stating that 'the property is sold free from legal incumbrances on completion' is simply referring to incumbrances on title or whether this is an express provision providing that vacant possession will be given on completion.

Case law has held that where a contract is silent as to vacant possession, and silent as to any tenancy to which the property is subject, there is impliedly a contract that vacant possession will be given on completion.[1] However, the implied assumption, that vacant possession is to be given, can be seen to be subject to specific

---

[1] *Cook* v *Taylor* [1942] Ch 349 at 352.

circumstances and actual knowledge of the parties. For example, where one party is aware, when entering into a contract, that the interest is subject to some impediment to vacant possession, case law would seem to suggest that if the purchaser knows (or is deemed to know) that the obstacle to the receipt of vacant possession is irremovable then the implied obligation to give vacant possession will not extend so as to include that obstacle. However, if at the time the contract was made, the purchaser knew (or was deemed to know) of only a removable obstacle,[2] then the implied obligation to give vacant possession under open contract rules will not be deemed to exclude such an obstacle, and if the removable obstacle is still on the premises on completion the obligation to procure vacant possession will have been breached.

It has been held that an express obligation to give vacant possession will prevail regardless of the nature of any known potential impediment to vacant possession (removable or irremovable).[3]

As such, if the clause stating that 'the property is sold free from legal incumbrances on completion' amounts to an express clause for vacant possession, then the items left behind will (subject to the *de minimis* threshold) cause there to be a breach of the obligation to give vacant possession. If not, and the obligation is merely implied, then because the items were known of by the parties at the time of contract, they will constitute a breach of the obligation if they are deemed to be removable. If they are irremovable (they are referred to as large items) then the seller will not be in breach and the purchaser will be required to take the property with the items thereon.

Unhelpfully, there is no actual guidance as to what is deemed removable or irremovable in this context. A lease has been said to be irremovable[4] but the determination will be a question of fact and degree in any given case.

Scenario 2

A seller contracts to convey land to a purchaser. The contract provides expressly that vacant possession is to be given on completion.

Between exchange and completion a third party asserts an adverse right to collect wood and other materials from the property that will prevent development of the land by the purchaser in the manner desired.

The third party successfully establishes such rights.

---

[2] *Norwich Union Life Insurance Society* v *Preston* [1957] 1 WLR 813 establishes that a purchaser's knowledge of a removable object to vacant possession is irrelevant. See also *Timmins* v *Moreland Street Property Co Ltd* [1958] Ch 110.

[3] *Sharneyford Supplies Ltd* v *Edge* [1987] Ch 305.

[4] *Hughes* v *Jones* [1861] 3 De GF&T 307. This is obviously subject to contra indications or other intentions of the parties as shown by the contract.

Whilst the purchaser may have contractual remedies against the seller with respect to disclosure of third party rights, from a vacant possession perspective, can the seller transfer the land to the purchaser on completion in compliance with the seller's obligation to give vacant possession?

The seller claims that the third party's right is clearly an interest over the land rather than a competing claim to possession, but the purchaser claims that it none-theless prevents delivery of the property free from a claim of right to the land (i.e. the right to pass and re-pass) that is adverse to the purchaser's interests.

Further, the purchaser may claim that the third party's right to pass and re-pass constitutes (albeit infrequent) third party occupation of the land in some way and argues that the adverse right is a legal impediment that prevents it from obtaining the quality of possession for which it had contracted.

ANALYSIS

This question addresses whether the legal impediment complained of (the adverse right) breaches the obligation to give vacant possession.

Compulsory purchase orders and requisitioning notices are the main types of 'legal impediments' to vacant possession. An analysis of these was undertaken in chapter 6 in the context of how they relate to claims to, and competing restrictions on, 'possession' of the property in question. It is, however, possible to acquire or be granted less extensive rights over land which *do not* amount to 'possession'.

There is no definition, as such, of so-called 'lesser interests', but such an expression is likely to refer to interests amounting to something short of exclusive possession. An example would be an incorporeal hereditament, which includes certain profits (for example, a non-possessory interest in land, which gives the holder the right to take natural resources such as petroleum, minerals, timber, or wild game from the land of another), or other 'rights over' rather than claims of 'possession to' land.

Whatever the type of profit (whether it be rights to graze stock, plant and harvest crops, quarry stone, sand or gravel, or take timber) in practice the exercise of a right gives the right holder a substantial degree of control over the burdened land.[5] As such, this scenario questions whether such rights, while amounting to less than possession but still encumbering the estate being transferred in some way, would also amount to being a legal obstacle to the receipt of vacant possession, if sufficiently substantial.

The difference between legal impediments like compulsory purchase orders and notices of requisition, and the potential legal impediment here, can be explained by reference to the nature of the right or interest. Unlike compulsory purchase orders

---

[5] Wonnacott, M *Possession of Land* (Cambridge, Cambridge University Press, 2006) p 141.

and requisitioning notices which pass the right to 'possession' of the property in question to the acquiring authority (or another party), or in the case of statutory restrictions on the user, prevent possession from being legally possible, so-called 'lesser interests' or 'rights over land' do not amount to barriers to 'possession' of the property, as they are only rights *over* the land, rather than competing claims to possession of the land. The right holder in the scenario posed is not claiming a right to exclusive possession (i.e. clearly not a freehold or leasehold interest) and the seller remains, at all times, the party with the right to possession of the property (to which the vacant possession obligation pertains), which it can transfer to the purchaser pursuant to a sale contract.

The case of *Horton* v *Kurzke*[6] establishes that the risk of a purchaser buying subject to an adverse lesser interest is a defect in *title*, thus making impediments which amount to *less* than possession not issues of vacant possession.[7] This marks such an issue out as distinct from the vacant possession obligation (which relates to competing claims to possession itself). This would seem logical; the scope and extent of an obligation to give vacant possession, dealing with barriers to 'possession', should not encompass *rights* which, by their very nature, do not amount to possession. Thus, legal impediments, in the form of compulsory purchase orders and requisitioning notices, can be distinguished from legal impediments such as certain profits and incorporeal hereditaments; the latter being legal rights amounting to less than possession of the land to which they pertain, and therefore not being relevant to the vacant possession obligation.

As such, the obligation to give vacant possession can be understood as being concerned with barriers to 'possession', but not all conceivable rights pertaining to the land which fall short of fully fledged possession. The decision in *Horton* therefore highlights the need for close analysis of the legal impediment complained of in order to determine the scope and extent of the obligation. So-called 'lesser-interests' are not issues of vacant possession but rather issues of title, and the right in question here would appear to fall into the non-vacant possession category.

SCENARIO 3

A seller contracts to convey a property to a purchaser. The contract provides expressly that vacant possession is to be given on completion.

The purchaser intends to grant a lease to a business tenant of the property on the same day. The tenant requires occupation that day given the nature of the tenant's business.

---

[6] *Horton* v *Kurzke* [1971] 1 WLR 769.
[7] See Megarry, W and Wade, W *The Law of Real Property* (London, Sweet and Maxwell, 7th ed, 2008) p 672.

On the morning of completion, the seller and purchaser complete and the purchaser is given the keys.

Later in the day the purchaser meets his proposed new tenant at the premises to sign the lease and hand the keys over. Upon inspection of the property the purchaser and the tenant see that certain items and materials have been left by the seller.

The tenant refuses to sign the lease because the tenant says that he cannot immediately occupy the property as he needs to. Instead, he takes a lease of an adjacent unit the following week. In two months time the purchaser manages to lease out the property to a third party at a rent lower than had been agreed with the tenant originally proposed due to a decline in the market.

The purchaser claims that the seller was in breach of his express contractual obligation to give vacant possession and claims that loss has been suffered as a consequence.

The seller claims that the items left were fixtures (and therefore part of the land) or, in the alternative, that the items and materials come within the *de minimis* threshold and therefore that no breach has arisen. Further, the seller states that leaving the items and materials behind was not inconsistent with him seeking to pass possession to the purchaser on completion, and that he did all he could to give vacant possession to the purchaser.

Analysis

In answering this question, it must first be ascertained whether the items are relevant to the obligation and, if so, whether they constitute a breach of the obligation to give vacant possession on the facts (with reference to the *de minimis* threshold). On the basis that a breach has occurred, the likely remedies available to the purchaser must then be considered.

Fixtures versus chattels

The proper determination of the status of the items can be seen as a preliminary issue in seeking to establish whether the items have been left behind by the tenant unlawfully and can therefore cause a breach of the vacant possession obligation.

It is common for disputes to arise as to whether items left behind at a property are fixtures (and therefore part of the land) or chattels (which are personal property of the tenant obliged to procure vacant possession, and which must therefore be removed). Indeed, it is commonly established that if the seller's failure to give vacant possession is due to the presence on the property of *chattels*, which affect usability of the premises, then a breach of the obligation to give vacant possession will arise if the impediment substantially interferes with enjoyment of a substantial part of the premises on completion.[8] This is why the distinction between fixtures and chattels has traditionally been seen to be so important.

Fixtures are physical objects which accede to the realty. Any physical object classed as a fixture as a matter of law merges with the land and title to it automatically vests in the owner of the freehold, and the object itself cannot be severed from the land by anyone other than the freehold owner.[9] Further, the purchaser of a freehold is entitled to all fixtures on the land at the date of exchange of contracts.[10] Chattels are physical objects which retain their independent character as personalty despite close association with realty. They thus do not attach to the land and do not pass with a conveyance of the land unless stipulated in the conveyance. Therefore, a seller is perfectly entitled, and indeed obliged, to remove such items before completion. This is all based on the maxim of law *quicquid plantatur solo, sols cedit*, meaning 'whatever is affixed to the soil accedes to the soil'.[11]

The fixtures and chattels distinction turns on two distinct but connected tests. The first test concerns the physical degree of annexation to the land. The more permanently and irreversibly the object is affixed to the land, the more likely it is to be considered a fixture. A form of gravity test for a chattel has developed out of this, in that an object that merely rests on the land due to its own weight will be classed as a chattel, and one more permanently fixed will be classed as a fixture. In *Holland* v *Hodgson*,[12] spinning looms bolted to the floor were classed as fixtures, but in *Hulme* v *Bingham*,[13] heavy machinery otherwise unattached was considered a chattel. In *Botham* v *TSB Bank Plc*,[14] kitchen appliances that were only connected electrically to the land (remaining in position by their own weight) were considered chattels on this test.

Gray and Gray[15] argue that the trend in case law suggests the above test is being overtaken by the second test concerning the objectively understood purpose (or object) of the annexation. The key question in respect of this test is whether the installation of the object was intended to effect a permanent improvement to the

---

[8] See *Cumberland Holdings Ltd* v *Ireland* [1946] KB 264. Also, Megarry and Wade, *The Law of Real Property* (2008) p 672 state that 'removable physical impediments' are relevant to the obligation — i.e. chattels and not fixtures which are attached permanently to the land and which pass under the contract of sale.

[9] A plethora of case law exists — see *Reynolds* v *Ashby & Son* [1904] AC 466 for example.

[10] *Taylor* v *Hamer* [2002] EWCA civ 1130.

[11] Burn, EH and Cartwright, J *Modern Law of Real Property* (Oxford, Oxford University Press, 17th ed, 2006) p 156.

[12] *Holland* v *Hodgson* [1872] LR 7 CP 328.

[13] *Hulme* v *Bingham* [1943] KB 152.

[14] *Botham* v *TSB Bank PLC* [1996] 73 P&CR D1, CA.

[15] Gray, K. and Gray, SF *Elements of Land Law* (Oxford, Oxford University Presss, 4th ed, 2006) p 39.

realty or was merely a temporary addition to the realty to enhance the enjoyment of the chattel.[16]

An example of the supposed importance of this distinction for vacant possession arose in the case of *Hynes* v *Vaughan*.[17] In this case, one issue surrounded a chrysanthemum growing frame and sprinkler system, and whether these could be argued to be fixtures or chattels. The seller defendants had removed these from the property after the date of the contract, which was unlawful if they were fixtures as they had passed with the land to the purchaser.[18] In view of the functions of the chrysanthemum growing frame and installation of the sprinkler system, it was determined that those items could not be seen as fixtures on the property so as to pass under the contract to the plaintiff. That is, the growing frame was not fixed to the land so as to pass to the purchaser under the contract. As such, the defendant sellers were correct to remove these items and, if they had not, this would have constituted a breach of the obligation to give vacant possession under the contract of sale (if sufficiently substantial).

Whilst the first question in this situation will be to establish whether the items were fixtures or chattels, with that said, it is relevant to note that case law can be seen to take a broad view that *any* impediment which prevented the purchaser from obtaining the quality of possession for which the purchaser had contracted would constitute a breach of the seller's obligation.[19] Particularly in the context of the impediment being more connected to the state and condition of the property, it must be borne in mind that the fixtures and chattels distinction may be somewhat artificial, with the relevant determination concerning whether the impediment complained of constitutes a substantial obstacle to the receipt of vacant possession on completion (irrespective of its status as a fixture or a chattel). As noted in chapter 7, there is currently no authority on the position where the vendor's inability to give vacant possession is due to the physical state of the property.[20] It is therefore arguable as to whether an impediment to vacant possession that is not a chattel, but more part and parcel of the state and condition of the property itself, can (in principle) amount to a breach of an obligation to give vacant possession, as the decision in *Hynes* appears to suggest.[21] As such, the scope of the obligation to give vacant possession may not just concern chattels, as has been traditionally perceived.[22] Potential obstacles connected to the state and condition of the property *could* also potentially be a barrier to vacant

---

[16] *Elitestone Ltd* v *Morris* [1997] 1 WLR 687, per Lord Lloyd at 690.

[17] *Hynes* v *Vaughan* [1985] 50 P. & C.R. 444.

[18] *Taylor* v *Hamer* [2002] EWCA civ 1130.

[19] *Korogluyan* v *Matheou* [1975] 239 EG 649.

[20] See Harpum, C 'Vacant Possession — Chamaeleon or Chimaera?' (1998) *Conveyancer and Property Lawyer* 324, 400 (CH).

[21] *Hynes* [1985] 50 P&CR 444.

[22] ibid, per Scot J at 453.

possession if they could be described as impediments which substantially interfere with the buyer's right to possession, and that the vacant possession tests must be applied in such circumstances. Case Study 3 addresses such issues in more detail.

Application to the tests

Assuming that the items left over *are* relevant to the vacant possession obligation, in seeking to ascertain whether there has been a breach of the obligation to give vacant possession, it would be necessary to apply the tests laid down in *Cumberland* and *John Laign*.[23]

The first limb of the tests is directed at the activities of the party required to give vacant possession (i.e. tenant on exercising a break option in a lease, or seller when transferring an interest in land, such as in this example) and provides that if the conduct of the party in question indicates that they, as seller or tenant, are continuing to use the premises for its own purposes in a non-trivial way (for example, by leaving goods in the premises), then it will fail to establish that vacant possession has been given. As such, this first limb focuses on the specific circumstances of the party required to give vacant possession and its conduct in so purporting to give vacant possession. This limb can therefore be seen to be inherently fact specific in nature and to refer to the actual seller or tenant in question, and its intentions, belief and state of mind as manifest by its conduct.

The second limb is directed at whether the contents of the premises present, objectively speaking, a *substantial* obstacle to the buyer's or landlord's own physical enjoyment of the premises on completion (or at the operative date when the obligation is engaged). If they do, vacant possession will not have been given.

On the facts, whether there is a breach of the obligation to give vacant possession here may depend on the nature of the seller's business and the items left over. If the seller is a builder, the buyer will have a strong argument for saying that by leaving, for example, building materials, in the property, the seller is still using the premises beneficially, for the storage of goods for the purposes of its business, and therefore that a clear intention has not been manifested to give vacant possession. As such, the seller will fail on the first limb of the *Cumberland test*.[24] Alternatively, if the materials were brought onto the property by the seller, which runs an office from the

---

[23] *Cumberland Consolidated Holdings Ltd* v *Ireland* [1946] KB 264 and *John Laign Construction Ltd* v *Amber Pass Ltd* [2004] 2 EGLR 28.

[24] This can be compared with *Legal & General Assurance Society Ltd* v *Expeditors International (UK) Ltd* [2006] EWHC 1008 (Ch); [2006] L&TR 22, where the judge decided that vacant possession had not been given because the warehouse was still being used for the storage of 'a few pallets and parcels in a largely empty warehouse'. It was noted that such items remained useful to the tenant's business. See also Fetherstonhaugh, G 'Can Premises that are Left Half Empty or Half Full be Vacant?' (2008) *Estates Gazette* 34.

premises, to repair the property but it did not complete this in time, different arguments may apply, and the buyer might succeed in establishing that the materials remaining in the premises on completion were not consistent with the seller continuing to use the property for its own purposes (the first limb), and (if not great in number) that such items did not constitute a substantial impediment to the buyer's right of possession on completion (the second limb). What the buyer, as the party with the right to vacant possession, will use the property for (compare a large industrial warehouse to a small corner shop) may also be key in determining whether the leftover items are a substantial impediment to the receipt of vacant possession. Indeed, the substantiality of the potential impediments is also something that must be considered in its own right.

De minimis

Whether the items can be argued to be *de minimis* would be a relevant point for determination.

De minimis is a Latin expression relating to 'minimal things', normally in the phrases *de minimis non curat praetor* or *de minimis non curat lex*, meaning that the law is not interested in trivial matters or that 'the law does not care about very small matters'.[25] The expression has also been used to describe a constituent or component part of a wider transaction, where it is in itself insignificant or immaterial to the transaction as a whole, and will have no legal relevance or bearing on the end result. In a more formal legal sense it means something that is unworthy of the law's attention. In risk assessment, for example, it refers to a level of risk that is too small to be concerned with; some refer to this as a 'virtually safe' level.[26]

Historically, it was unclear as to whether a *de minimis* threshold operated in determining whether the obligation to give vacant possession had been breached, and this may explain the differing decisions reached by respective judges on ostensibly similar questions of fact.[27] More recently, case law has suggested that the obligation is qualified as being subject to a *de minimis* rule.[28] Lord Greene in *Cumberland Consolidated Holdings Ltd* v *Ireland* stated:

> Subject to the rule de minimis *a vendor who leaves property of his own on the premises on completion cannot, in our opinion, be said to give vacant possession.*[29]

---

[25] See Ehrlich, E *Amo, Amas, Amat and More* (New York, Harper Row, 1985) p 100. It literally means that 'the law does not concern itself with trifles'.

[26] See the National Library of Medicine Toxicology Glossary — Risk *De minimis*.

[27] For example, in the cases of *Savage* v *Dent* [1736] 2 Stra 1064 and *Isaacs* v *Diamond* [1880] WN 75.

[28] *Cumberland* [1946] KB 264.

[29] ibid, per Lord Greene at 270. Emphasis added.

In the context of vacant possession, *de minimis* can be seen to refer to small or insignificant obstacles to the receipt of vacant possession. Whilst case law indicates that the vacant possession obligation will be subject to a *de minimis* rule,[30] how that operates in practice and what such a threshold refers to remains unclear, however, and has not been elaborated upon by the courts. For example, in *Legal & General Assurance Society Ltd* v *Expeditors International (UK) Ltd*,[31] a rubbish bin, a table, coffee mugs and a swivel chair left at the premises were considered unimportant and justified no further reference in the decision on the point. Clearly, by themselves, the items would not have prevented vacant possession from being given. By contrast, in *Cumberland*[32] rubbish that filled two-thirds of the warehouse cellars led the Court to hold that vacant possession had not been given.

It is difficult to draw the line when the facts lie somewhere between these two examples. What is clear, from the decision in *Cumberland*, is that the interference must be of some substantial nature:

> *When we speak of a physical impediment we do not mean that any physical impediment will do. It must be an impediment which substantially prevents or interferes with the enjoyment of the right of possession of a substantial part of the property.*[33]

As such, it is clear that an element of substantiality is manifest in the determination as to whether the party with the right to vacant possession can commence occupation without difficulty or objection. This would be a question of fact with reference to the impediment complained of. The issue of *de minimis* is also relevant to interpreting the first limb of the test, and seeking to ascertain whether what is left behind by the party seeking to give vacant possession is significant enough for it to be argued that the non-removal of such items is inconsistent with its attempts to deliver vacant possession, and more consistent with it seeking to continue to use the premises for its own purposes (namely the storage of goods).

Damages

On the basis that the seller is found to be in breach of its obligation to give vacant possession, the purchaser is likely to sue the seller for damages as a result of not giving vacant possession on completion. It is unlikely in this case, given that

---

[30] Following *Cumberland* [1946] KB 264 where the obligation was stated as being subject to such a rule.

[31] *Legal & General Assurance Society Ltd* v *Expeditors International (UK) Ltd* [2006] EWHC 1008 (Ch); [2006] L&TR 22. First instance decision. Upheld on appeal see *Legal and General Assurance Society Ltd* v *Expeditors International (UK) Ltd* [2007] All ER (D) 166 (Jan).

[32] *Cumberland* [1946] KB 264.

[33] ibid, per Lord Greene at 287.

completion has taken effect and the purchaser has let the premises to an alternative tenant, that the purchaser would seek to action a breach of the obligation by purporting to rescind the contract on the grounds of non-merger. It would not be possible in this instance to serve a notice to complete, or seek specific performance.

The availability and quantum of damages will be determined by the circumstances and the nature of the losses in question. A purchaser's remedies may also be restricted by the express terms of the contract.[34] However, as a general rule, a purchaser is likely to be able to recover as damages the sum necessary to place it in the position it would have been in if the contract had been performed.

Whether an impediment is removable or irremovable will be a question of fact and degree. Where the impediment is removable (as appears to be the case in this example) then the purchaser may recover the cost of actually removing the impediment to vacant possession. In *Cumberland*[35] the purchaser was successful in the recovering the cost of removing rubbish left in the premises. Further, following *Beard* v *Porter*,[36] the purchaser will *also* seek to obtain consequential damages with reference to the reduced rental income obtained from the alternative letting, and other consequent effects. If the impediment was said to be irremovable, then the measure of damages will be the difference between the purchase price and the market price of the property subject to the impediment.

Case study 2: vacant possession in the leasehold context

This case study covers issues of vacant possession in the leasehold context.

SCENARIO

A tenant has trouble paying the rent under a lease and decides to move to smaller premises. The tenant purports to exercise a break option in its lease with the landlord which provides:

> *Right to Terminate*
> *The Tenant may terminate this lease by giving not less than 6 month's prior written notice, and provided that the Tenant has complied with all covenants in the Lease up to and including the termination date as specified in the Tenant's notice, then the Lease will terminate without prejudice to antecedent claims and liabilities.*

---

[34] See PLC Property, 'Selling with Vacant Possession' (accessible via subscriber service).
[35] *Cumberland* [1946] KB 264 at 270.
[36] *Beard* v *Porter* [1948] 1 KB 321.

The lease also included the following clause:

Yielding-up
The Tenant will yield-up the Demised Premises to the Landlord upon termination of the Lease.

The landlord accepts the notice as valid. The break will therefore take effect subject to compliance with conditionality.

On the break date the premises are not cleared of the tenant's chattels. The tenant claims that vacant possession was not a condition of the break option in the lease, and as such the lease has ended. The landlord disagrees.

The tenant is concerned that the lease does not continue, as a rent review is due the following month.

QUESTIONS

1. Is the break conditional on vacant possession?
2. How would the answer to (1) change if:
   a. the break was conditional upon 'compliance with all covenants in the lease immediately prior to the break date'?
   b. the break was conditional upon 'reasonable compliance with all covenants in the lease up to and including the break date'?
   c. the yielding-up clause stated that 'the tenant will yield-up the demised premises upon expiry of the lease'?
   d. the lease contained no express yielding-up clause?
3. In respect of seeking to comply with the yielding-up obligation in the lease, what should the tenant do in connection with such matters as (a) partitioning (b) electrical wiring and (c) new fixtures which it has added during the term of the lease?
4. On the basis that the lease continues past the break date:
   a. what assumption will apply on rent review as to vacant possession in the absence of any express provisions?
   b. will a sublease of the premises have any bearing on the valuation at rent review?
   c. why is this important to the tenant?

QUESTION ANALYSIS

1. The landlord will take the position that the condition which states that the tenant must have complied with all covenants up to and including the break date includes the tenant's covenant to yield-up at the end of the lease, which includes the return of the premises with vacant possession (subject to certain exceptions, for example in

connection with subtenants who have statutory protection). The landlord will claim that the chattels remaining in the property on completion is inconsistent with the yielding-up obligation having been complied with.

In this regard, and as highlighted in chapter 1, a landlord can use the issue of vacant possession to prevent the tenant exercising a contractual break option in a lease if the landlord would prefer the lease to continue. This is an example of the implied requirement to give vacant possession arising as a consequence of the yielding-up obligation in the lease being incorporated as a condition of the break; this is by virtue of the requirement that the tenant must have complied with all covenants as at the break date. In such as case, it is likely that a tenant will not appreciate this until the issue is raised, normally *after* the break date, when the tenant's opportunity to give vacant possession will have passed.

If the landlord is correct, the tenant will remain liable under the terms of the lease, which will continue.

2. The answers would change in the following ways:

2(a): If a break is conditional on compliance with tenant covenants 'up to the break date' or 'immediately prior to the break date', and not 'up to and including the break date' or 'at the break date', then the conditionality will not refer to covenants (express or implied) which only take effect 'on' the break date itself. The yielding-up obligation (including the requirement to vacate) arises (expressly or impliedly) on the last day of the term (i.e. in this case, the break date) but not before. As such, if the break was conditional upon 'compliance with all covenants in the lease immediately prior to the break date', then the yielding-up obligation would not be part of the conditionality associated with the break exercise, and the leftover chattels would *not* prevent the lease determining on the basis of not yielding-up the property on the break date. The tenant would have to ensure that all other relevant covenants immediately prior to the break date had been complied with, however, in order to operate the break.

2(b): If the break was conditional upon 'reasonable compliance with all covenants in the lease up to and including the break date', the tenant may argue that to comply with the obligation to yield-up (which, as a covenant in the lease, would become part of the conditionality of exercise of the break) only a reasonable attempt to yield-up (a lesser obligation) would be sufficient to operate the break option in such an instance, and that a more generous *de minimis* threshold would be applied to the left over chattels in the property. In this case, the conditionality may be seen to be qualified by such wording, unlike an express pre-condition to give vacant possession which would be an absolute condition when appearing in its own right.

2(c): Here, the actual wording of the yielding-up obligation will be crucial. The normal phrase used in a yielding-up obligation is 'on the expiry or sooner determination of the term' but if there is reference only to 'expiry' then compliance may not be required on earlier termination pursuant to a break option. This would be determined by an interpretation and construction of the terms of the document in context.

The tenant would argue that the requirement to give vacant possession is part of the yielding-up obligation which only arises on expiry of the lease, but that it is not part of the pre-conditionality associated with the earlier termination pursuant to the break right.

2(d): According to *Henderson* v *Squire*,[37] there is an implied term in a lease that the tenant will deliver the demised premises back to the landlord at the end of the term in the absence of an express clause:

> *The question is, where there is a tenancy, and nothing is expressed as to delivering up possession at its determination, whether there is an implied contract that the tenant shall not only go out of possession, but restore the possession to the landlord...I think that there is such an implied contract.*[38]

Subject to any express or contrary indications, there is still therefore a covenant to yield-up at the end of the term.[39] If the break is conditional upon compliance with all covenants in the lease up to and including the break date, that will include the implied covenant to yield-up the premises on the break date (subject to the break constituting the 'end of the term', based on a proper construction of the document).

3. As noted in previous chapters, the state and condition of a given premises is relevant to the procurement of vacant possession, and therefore a matter which must be considered carefully upon the exercise of breaks which are conditional on either vacant possession, yielding-up, or repair obligations more generally.

The state and condition that the premises must be returned in will be determined by the terms of the lease (including, in particular, obligations relating to reinstatement) and also supplemental licences and documents pertaining to alterations. The terms of these will be crucial in seeking to establish whether certain 'items' in the premises can stay, or must go.

The cost of reinstating, or stripping out, a property and complying with repairing obligations under the lease, and then making good any damage caused by the removal of such structures from the premises, can be excessive. As such, the extent of a tenant's obligations will commonly be a source of dispute and uncertainty. A tenant will want to spend the minimum possible to comply with its obligation to give vacant possession (and any other tenants' covenants upon which the break is conditional), but at the same time will not want to fall foul of the conditionality, thus causing the lease to continue.

A yielding-up clause may expressly require the tenant to remove fixtures and any alterations it has made to the property, in addition to making good damage caused in connection with their removal. In relation to any dispute about the meaning of any

---

[37] *Henderson* v *Squire* [1869] LR 4 QB 170.

[38] ibid, per Blackburn J. See also the judgment in *Harding* v *Crethorn* [1973] 1 Esp 56.

[39] See *Hyatt* v *Griffiths* [1981] 17 QB 505 and *Newson* v *Smythies* [1858] 1 F&F 477.

express covenant to remove fixtures, the covenant will be interpreted using the normal principles governing interpretation of contracts and other legal documents.

Quite commonly, so-called 'strip-out and reinstatement' obligations will not be included in a yielding-up clause in a lease however. This is because a reinstatement obligation may be viewed as onerous at rent review, if the reinstatement obligation is not specifically referred to as a 'disregard'. Further, a landlord may not actually want the tenant to reinstate if, for example, the property is to be demolished, and may not want such a default provision to be provided for by the lease. If no reinstatement provision is included at all in the yielding-up clause in relation to reinstatement (as would appear to be the case here), then the tenant's obligations will be entirely determined by the terms of the lease and any consents granted for alterations, or individual enforcement measures.

As a general rule, the tenant is not obliged to remove things that are 'part of the building' or that are landlord's fixtures. The tenant is also not obliged to remove tenant's fixtures, although it is entitled to do so until the last minute of the term.[40] Often, leases state that tenants must remove demountable partitioning but in the absence of clear words, removal of all partitioning may be required (or advisable). A covenant to leave premises 'in good repair and condition' was held to require the tenant to remove partitioning in a case from the Australian courts, but it is unclear to what extent this will be followed in England and Wales.[41]

For a tenant, it is always advisable, where possible, to seek to agree reinstatement obligations with the landlord in advance, but often a landlord may wish to be unhelpful in this regard. In such cases, a tenant should consider agreeing to surrender the lease (with payment of associated premium if necessary) in order to risk non-compliance with the conditionality associated with the break.

4. The respective points are relevant on rent review:

4(a): Rent review clauses will usually provide that the property will be valued on the basis that it is being let with vacant possession. As such, the tenant will be treated as having moved out (or as never having occupied the premises), and the rent will be calculated on the assumption that the tenant has removed its tenant's fixtures and chattels from the premises (and that the premises are ready for occupation by a tenant).

If there is no express provision in the lease stating that the premises are to be valued on the basis that they are to be let with vacant possession (as would seem the case here), the normal assumption is that a lease should to be valued on such a basis (subject to express contrary wording or indications). It would therefore be necessary to review the terms of the lease, and the rent review clause, carefully to check for any contrary indications.

---

[40] *Never-stop Railway (Wembley) Ltd* v *British Empire Exhibition (1924) Inc* [1926] Ch 877.

[41] See *Wincant Pty Ltd* v *State of South Australia* [1997] 69 SASR 126.

4(b): A particular issue relates to subleases. A lawful sublease (or licence) may be taken into account in the rent review valuation in appropriate cases, with the consequent effect on valuation. As such, it would be necessary to determine whether the lease was granted specifically subject to an existing tenancy, or whether the lease clearly contemplated the creation of an underlease. Case law indicates that all relevant circumstances must be taken into account in such instances in determining whether the valuation is to have regard to an underlease of the whole or part of the premises.

If an underlease was granted in breach of covenant, then the subletting should be ignored in deciding whether vacant possession should be assumed. Otherwise, the effect of taking into account the subletting would likely be to decrease the rent, unfairly allowing the tenant to benefit from its own wrongdoing by acting in breach of covenant under the lease and granting a sublease (not contemplated by the parties to the lease).

4(c): In terms of why a disregard of occupation, and assumption of vacant possession, will be important to the tenant, when a landlord and tenant (who is already in occupation) are negotiating a new rent, the fact that the tenant is already in occupation could potentially induce it to offer a sum greater than it would offer if it were not in occupation of the demised premises that are subject to the review. This is logical when one considers that the tenant in occupation is likely to view the expense and disruption to its business of having to vacate and relocate to other premises as having an intrinsically greater value, which would not be apparent with respect to a third party tenant moving to the premises from elsewhere. The tenant, in occupation, will likely attach a value to remaining in the current premises compared to having to move elsewhere.

As such, the mischief behind the disregard of occupation, and assumption of vacant possession, is to prevent the landlord benefitting from an increase in the rent for the premises attributable to the fact of occupation by the tenant itself (and, where appropriate, any predecessor in title).[42] The tenant would also seek to obtain a disregard of goodwill in its own right, as otherwise this could have the effect of increasing the likely rent payable for the ongoing tenancy.

Section 34 of the Landlord and Tenant Act 1954 directs the court to disregard (*inter alia*) certain matters when determining the rent that is to be payable for a new business tenancy. It is the disregards contained in this section which are often referred to in standard rent review clauses in leases to specifically address the issues that are referred to above. In respect of vacant possession, two disregards are of importance, namely:

- any effect on rent of the fact that the tenant has or its predecessors in title have been in occupation of the holding; and

---

[42] See *East Coast Amusement Co* v *British Transport Board* [1965] AC 58 and *Hambros Bank Executor & Trustee Co* v *Superdrug Stores* [1985] 1 EGLR 99.

- any goodwill attached to the holding by reason of the carrying on thereat of the business of the tenant (whether by the tenant or by a predecessor of the tenant in that business).[43]

A tenant would be well advised to ensure that these (or equivalent provisions) are expressly inserted in the rent review clause in a lease.

Case study 3: vacant possession and the state and condition of the property

This case study addresses issues relevant to the state and condition of the property on completion, and their impact on an agreement which provides for vacant possession.

CHRONOLOGY

In September the landlord and tenant agreed that upon completion of certain works (defined in the 'Agreement' as 'Landlord's Works') the landlord would grant the tenant a lease of part of the building (the 'Demised Premises') for use as office premises.

The Landlord's Works involved the removal/replacement of internal partitions; the creation of independent power, lighting, data and telecoms services; and redecoration and re-carpeting.

In October, the tenant began its fitting-out works.

In November, the parties' surveyors agreed floor areas – which fixed the completion date as 17 November.

On 3 November, the landlord's solicitors sent the engrossed lease to the tenant's solicitors for execution by the tenant. This was returned on 7 November.

On 17 November the new lease should have been completed but it was not.

On 11 December, an explosion occurred damaging and/or destroying the premises. The cause of the explosion has not been ascertained.

In January the following year the tenant wrote to the landlord stating that the tenant would complete the lease, provided the landlord confirmed that it would honour the nine month rent free period following completion of the rebuilding works.

In February, the landlord's solicitors sought confirmation that the tenant's solicitors were holding the executed lease and were prepared to complete.

Completion was never effected.

---

[43] Landlord and Tenant Act 1954 s 34(1)(a) and (b).

THE AGREEMENT

The relevant provisions of the agreement (and the lease) were as follows:

Lease Completion Date
The date 12 working days after ...the date of the Agreement becoming unconditional

Vacant Possession
On completion of the Lease vacant possession of the Demised Premises shall be given to the Tenant

Right to Terminate
Either party may determine this Agreement forthwith by giving notice to the other to that effect if that other party shall have committed any material breach of its obligations under this Agreement ...

Insurance
The tenant covenanted to pay '... a proportionate part of the yearly sum which the Landlord shall from time to time pay by way of premium or premiums for:

(a) *keeping the Building insured ... against the Insured Risks in the full reinstatement value thereof and in the event of damage or destruction by any of the Insured Risks and when lawful to do so to cause all monies received by virtue of such insurance (other than in respect of the loss of Rent) to be laid out in rebuilding or reinstating (so far as practicable) the Building*

(b) *insuring loss of the Rent payable under this Lease ... for three years ...'*

Cesser of Rent
If the Building or any part thereof shall at any time be destroyed or damaged by any of the Insured Risks so as to render the Demised Premises unfit for occupation and use ... or inaccessible then the Rent ... and the Service Charge shall be suspended and cease to be payable until the Demised Premises shall again be rendered fit for occupation and use ...

Determination following Destruction
If three months prior to the end of the loss of rent period for which the Landlord insures, the Demised Premises or access to it are not fit for occupation and use by the Tenant either party may by service of not less than three months written notice determine the Term ...

Term
10 years from contractual completion with a nine month rent free period from that date.

User
Use as offices.

QUESTIONS

1. What obligation to give vacant possession arose in this case?
2. How can the tenant argue that vacant possession cannot be given in this case?
3. How can the landlord argue that vacant possession can be given in this case?
4. Is it possible to claim that the obligation to give vacant possession has been waived?
5. Could the tenant terminate the agreement under the contractual termination provision contained in the agreement?
6. Could it be argued that there has been a frustration of the agreement for lease?

QUESTION ANALYSIS

The nature of the landlord's obligation to confer vacant possession

The clause contains an express contractual provision for vacant possession. The question then is whether it can be said that 'vacant possession' can be given in circumstances where the proposed premises forming the demise are the subject of extensive damage.

There is no authority specifically on whether the state and condition of a given property can be a barrier to vacant possession.[44] The closest authority on point is that of *Topfell* v *Galley Properties Ltd.*[45] In that case, a company had agreed to sell premises subject to a tenancy of the first floor but with vacant possession of the ground floor. It was held that the landlord was in breach of contract in failing to give vacant possession when a notice had been served by the council which limited occupation of the property to one household only. Templeman J said that:

> The vendors say that all that vacant possession of the ground floor means is that it is empty and that, physically, there is nothing to stop the purchasers from entering. The purchasers, on their part, say that vacant possession of the ground floor is a contractual requirement on the vendors which binds the vendors to deliver the property in a state in which it can be enjoyed. Vacant possession, they say, is the right to occupy and enjoy the property either by the purchaser himself or by his tenants or licensees.
>
> In my judgment the purchasers are right about this. The meaning of the words 'vacant possession' can, I think, vary from context to context but the background to this case is that to all outward appearance, the house consisted

---

[44] See Harpum, 'Vacant Possession – Chamaeleon or Chimaera?' (1998) *Conveyancer and Property Lawyer* 324, 400.

[45] *Topfell* v *Galley Properties Ltd* [1979] 2 All ER 388.

of two separate occupations…In my judgment, when the vendors said they would give vacant possession in the context of this property, the vendors cannot now say, 'Oh no; all we intended and all we contracted to give was the right to possession in the negative sense. There is no rubbish on the floor, no other tenants and nobody else was there. It was vacant.' I have come to the conclusion that [the sellers] were contractually bound, on completion, to hand over the ground floor in a condition which would allow the plaintiffs to occupy it. It is quite plain that at the date of the contract and at the date fixed for completion, the vendors cannot do that because, by reason of the Housing Act direction, in fact, nobody can occupy the ground floor. The vendors cannot occupy it themselves, they cannot sell it to somebody who wishes to purchase it in order to go and live there himself and they cannot let it.[46]*

Three relevant principles can be extracted from this case, namely:

- The meaning of 'vacant possession' may vary from context to context. In undertaking the construction process, the Court's task would be to ascertain the mutual intentions of the parties with respect to the legal obligations assumed by the contractual words in which they were sought to be expressed. The intention of the parties would have to be ascertained from the language which they had used in the agreement, considered in the light of the surrounding circumstances and the object of the agreement. The Court would not, however, embark upon the task of ascertaining the actual intentions of the parties. For example, the parties could not give evidence as to what their intentions really were. The Court would be concerned to ascertain, not what the actual intention of the actual parties to the lease had been, but what would have been the intention of the hypothetical reasonable parties to the lease, placed in the same position as the actual parties and contracting in the words used by the actual parties.
- The delivery of an empty property will *not* always be sufficient to amount to the grant of vacant possession.
- The existence of a notice preventing occupation of a property may prevent a landlord from being able to grant vacant possession.

The tenant's argument that vacant possession cannot be granted

In the light of the *Topfell* decision, there would be an argument in this case that as the agreement was for the grant of a lease with a specified user of 'offices', the landlord's obligation to grant vacant possession on completion carried with it an obligation to

---

[46] ibid, per Templemen J at 392.

deliver premises which were capable of being used and enjoyed as offices. The strength of this argument would be improved further if:

- Any statutory notices were in existence which precluded anyone from occupying or using the premises.
- The inability to use or enjoy the premises was prohibited for a very considerable period of time such as three years, either by virtue of notices or by virtue of the impracticability of effecting full remediation works within that time.

The landlord's position that vacant possession has been given

In addition to an argument that the obligation to grant vacant possession is satisfied by the delivery of empty premises, the landlord would argue that if the lease was completed, then the lease has provisions which cater for the situation which has occurred and hence, the lease could be completed and those provisions be allowed to take effect. In particular, reliance might be placed upon the fact that there is a rent cesser provision in the lease which would have the effect that rent would not be payable in respect of any period for which the premises were unfit for use and occupation or inaccessible due to an insured risk.

There are problems with such an approach however:

- It is no answer to an inability to comply with the obligation to grant vacant possession at *the moment of completion* to point to obligations which are intended to protect a tenant's position in the event of damage or destruction during the term of a lease. As noted in chapter 5, the obligation to give vacant possession arises *at the point* of completion.
- The rent cesser provisions only operate in the event of destruction or damage 'by any of the Insured Risks' and there may be uncertainty on the question of whether the damage has been caused by an insured risk. This would be a fact specific determination.

Has there been a waiver of the right to vacant possession?

It is possible that the landlord might also argue that there has been a waiver of the tenant's right to vacant possession or the tenant has affirmed the agreement. This argument might be based on a contention that the tenant had indicated a willingness to proceed notwithstanding their awareness of the argument that vacant possession could not be delivered upon completion. As noted in the scenario, in January the following year, the tenant wrote to the landlord stating that the tenant would complete the lease provided the landlord confirmed that it would honour the nine month rent free period following completion of the rebuilding works.

This reflects a need to preserve any rights in relation to a vacant possession obligation and ensure correspondence, following a completion date on which it is argued that vacant possession has not been given, is without prejudice to such a contention.

Could the tenant terminate the agreement under the contractual termination provision?

The other approach which could be open to the tenant would be to take steps to terminate the agreement under the contractual termination provision on the grounds that the landlord has committed a 'material' breach of its obligations.

It is true that if the landlord is unable to deliver vacant possession upon completion, then it is likely that this would be a breach of the agreement. The question which then arises is whether it can be said to be a 'material breach'. This would be a question of fact and degree. Time would have to be of the essence of the contract in order to rescind on such a basis, otherwise there would be the need to serve a Notice to Complete prior to rescission.

This reflects the need to check a contract more generally for an alternative (perhaps contractual) basis on which rescission could take effect.

Could it be argued that there has been a frustration of the agreement for lease?

The doctrine of frustration applies where there supervenes an event (without default of either party and for which the contract makes no sufficient provision) which so significantly changes the nature (not merely the expense or onerousness) of the outstanding contractual rights and/or obligations from what the parties could reasonably have contemplated at the time of its execution that it would be unjust to hold them to the literal sense of its stipulations in the new circumstances. In such a case the law declares that both parties be discharged from further performance.[47] For example, a lease may be frustrated where there has been some 'vast convulsion of nature' which has caused the destruction of the subject matter of the lease[48] or where, in relation to a short term lease of a building or structure, it is destroyed[49] or burns down.[50] However, a lease will not be frustrated if, for example, three and a half years of a 10-year term remain and access is closed for merely a year[51] or if a tenant is ousted from the premises by government requisition.[52]

---

[47] See Lord Simon of Glaisdale in the case of *National Carriers* v *Panalpina (Northern)* [1981] AC 675.

[48] *Cricklewood Property Investment Co* v *Leightons Investment Trust* [1945] AC 221.

[49] *National Carriers* v *Panalpina (Northern)* [1981] AC 675 which referred to a lease of an oil storage tank which was destroyed by an explosion.

[50] *Taylor* v *Caldwell* [1863] 3 B&S 826.

[51] *National Carriers* [1981] AC 675.

[52] *Matthey* v *Curling* [1922] AC 180 and *Whitehall Court* v *Ettlinger* [1920] 1 KB 680.

In this case, the first problem with the application of this doctrine is that although the doctrine can, in theory, apply to an agreement for lease,[53] it seems that it will be extremely rare that it will do so. This seems to be partly as a result of the principle that upon the making of the agreement, the proposed tenant is regarded in equity as the tenant and hence is bound to complete.[54]

The second problem with the application of the doctrine is that if the agreement makes provision for the express event which has occurred then this will normally preclude the application of the doctrine.[55] It could be said in this case that the rent cesser provision constitutes an express provision dealing with situations of damage or destruction, with the consequence that the parties are to be taken to have agreed that the loss will not fall on the tenant if the damage is as a result of an insured risk but that it will fall on the tenant if the damage is not as a result of an insured risk.

The third problem is that it is possible that, for example, the damage to the premises could be put right in, say, one to two years such that the premises could subsequently be occupied for the following eight years of the term. In that event, this situation would not be one which 'so significantly changes the nature (not merely the expense or onerousness) of the ...obligations from what the parties could reasonably have contemplated at the time' of the execution of the agreement. As discussed in chapter 6, frustration is rarely claimed for this reason.

---

[53] *Denny Mott & Dickson* v *James B Fraser & Co* [1944] AC 265 and *Rom Securities* v *Rogers* [1967] 205 EG 427.

[54] See *Chitty on Contracts* (London, Sweet & Maxwell, 25th ed, 1983) para 23-055.

[55] ibid.

Table of Cases

Table of Legislation

Index

Printed in the United States
by Baker & Taylor Publisher Services